Visioning Technologies

Visioning Technologies brings together a collection of texts from leading theorists to examine how architecture has been, and is, reframed and restructured by the visual and theoretical methods introduced by different 'technologies of sight' – understood to include orthographic projection, perspective drawing, telescopic devices, photography, film and computer visualization, amongst others.

Each chapter deals with its own area and historical period of expertise, organized sequentially to mark out and analyse the historical evolution of how architecture has been transformed by technologically induced shifts in human perception from the 15th century until today. This book underlines the way in which architectural forms and design processes have developed historically in conjunction with the systems of sight we manufacture technologically and suggests this continues today. Paradoxically, it is premised on the argument that these technological systems tend, in their initial formulations, to obtain ever greater realism in our visualizations of the physical world.

Graham Cairns is an academic and author in the field of architecture who has written extensively on film, advertising and political communication. He has held Visiting Professor positions at universities in Spain, the UK, Mexico, the Gambia, South Africa and the US. He has worked in architectural studios in London and Hong Kong and previously founded and ran a performing arts organization, Hybrid Artworks, specializing in video installation and performance writing.

He is author and editor of multiple books and articles on architecture as both a form of visual culture and a socio-political construct. He developed this book during a two-year period as Visiting Scholar at Columbia University, New York. He is currently Director of the academic research organization AMPS (Architecture, Media, Politics, Society), and Executive Editor of its associated journal *Architecture_MPS*. He is Honorary Senior Research Associate at the Bartlett School of Architecture, University College, London, UK.

Visioning Technologies
The Architectures of Sight

Edited by Graham Cairns

LONDON AND NEW YORK

First published 2017
by Routledge

2 Park Square, Milton Park, Abingdon, Oxfordshire OX14 4RN
711 Third Avenue, New York, NY 10017

Routledge is an imprint of the Taylor & Francis Group, an informa business

British Library Cataloguing in Publication Data
A catalogue record for this book is available from the British Library

Library of Congress Cataloging in Publication Data
Names: Cairns, Graham, author.
Title: Visioning technologies : the architectures of sight / Graham Cairns.
Description: New York : Routledge, 2017. | Includes bibliographical
    references and index.
Identifiers: LCCN 2016026538 | ISBN 9781472454966 (hardback : alk. paper)
Subjects: LCSH: Architecture and technology. | Image (Philosophy) |
    Technology–Social aspects.
Classification: LCC NA2543.T43 C35 2017 | DDC 720/.47–dc23
LC record available at https://lccn.loc.gov/2016026538

ISBN: 978-1-4724-5496-6 (hbk)
ISBN: 978-1-138-61668-4 (pbk)

Typeset in Sabon
by Apex CoVantage, LLC

Contents

Figures

Contributors

Iñaki Bergera received his degree in architecture (1997) and PhD (2002) from the University of Navarra, ETSAUN. He earned a master's from Harvard University GSD (2002). His PhD dissertation was awarded and published by the Fundación Arquia (Barcelona, 2005). He is Associate Professor at the School of Engineering and Architecture of the University of Zaragoza, having taught formerly at the ETSAUN (1997–2007) and at the Universidad Europea in Madrid (2007–09). He has been visiting teacher and guest critic in several international universities. He is involved in photography both as practice and as scholarly endeavor, and he understands architecture and photography as interdisciplinary parallel subjects. He has widely published and lectured on the topics and conducted specific research at the CCA in Montreal (2015 and 2010), the Getty Center in Los Angeles, the CCP in Tucson, Arizona, Columbia University GSAPP and ICP in New York (2012) and the University of Zurich (2014). He is the Principal of the Research Project *Photography and Modern Architecture in Spain, 1925–1965* (2013–2016) and curator of the homonymous exhibition held at the Museo ICO in Madrid (2014). He has received several international architecture awards.

Graham Cairns has written extensively on film, advertising and political communication. He has held Visiting Professor positions at universities in Spain, the United Kingdom, Mexico, the Gambia, South Africa and the United States. He has worked in architectural studios in London and Hong Kong and previously founded and ran a performing arts organization, Hybrid Artworks, specializing in video installation and performance writing. He is author and editor of multiple books and articles on architecture as both a form of visual culture and a socio-political construct. He developed this book during a two-year period as Visiting Scholar at Columbia University, New York. He is currently Director of the academic research organization AMPS (Architecture, Media, Politics, Society), and Executive Editor of its associated journal *Architecture_MPS*. He is Honorary Senior Research Associate at the Bartlett School of Architecture, University College, London, United Kingdom.

Valeria Carullo moved to London after earning a master's degree in architecture in Italy to work in the RIBA Library Photographs Collection and to study photography. After a few years assisting architectural photographer Richard Bryant, she returned to the RIBA where she has been Curator of photographs since 2012. Valeria has written several articles and given numerous talks on both architectural and photographic subjects. She has co-curated two major exhibitions, Framing

Modernism: Architecture and Photography in Italy 1929–1965 (Estorick Collection of Modern Italian Art, 2009, and MAXXI, 2011) and Ordinary Beauty: The Photography of Edwin Smith (RIBA Architecture Gallery, 2014); curated Art Deco Triumphant: The Exposition Internationale des Arts Decoratifs et Industriels Modernes, Paris 1925 (RIBA, 2011) and John Pantlin: Photographing the Mid-Century Home (Geffrye Museum, 2014) and contributed to several other exhibitions at the RIBA. In 2014, she organized the RIBA international symposium Building with Light: The Legacy of Robert Elwall on the relationship between photography and architecture. The proceedings of the conference will be published in late 2016 in the *Journal of Architecture* and will include Valeria's paper 'Image Makers of British Modernism: Dell & Wainwright at the *Architectural Review*'.

Nader El-Bizri is Professor of Philosophy and Director of the Civilization Studies Program at the American University of Beirut. He was previously a Principal Lecturer/Reader in Architecture at the University of Lincoln, a Lecturer in Architecture at the University of Nottingham and an Affiliated Lecturer in History and Philosophy of Science at the University of Cambridge. He also acted as a Senior Researcher at the Institute of Ismaili Studies in London and at the Centre National de la Recherche Scientifique in Paris. He holds a PhD in philosophy from the New School for Social Research in New York and an M.Arch-II in architecture from the GSD at Harvard University. He published widely in architecture, history of science and philosophy and serves on the editorial boards of publications by Oxford University Press, Cambridge University Press, Springer and E. J. Brill. He was also a consultant to the Science Museum in London, the Solomon Guggenheim Museum in New York/Berlin, the Aga Khan Trust for Culture in Geneva and contributed to BBC TV/Radio cultural programs. Besides his service to academia, he practiced as an architect in offices in New York, Cambridge, London and Beirut.

Federica Goffi is Associate Professor (2007–present) and Associate Director of Graduate Programs at the Azrieli School of Architecture and Urbanism at Carleton University in Ottawa, Canada. She is currently teaching architectural drawing, studio and a PhD colloquium. She has been an Assistant Professor at the Interior Architecture Department of the Rhode Island School of Design (2005–2007). She holds a PhD from the Washington-Alexandria Architectural Center of the Virginia Polytechnic Institute and State University. She has published book chapters (Ashgate, Routledge) and journal articles on the threefold nature of time-weather-tempo, as it informs notions of 'built conservation' (AD, ARQ, In.Form, Interstices, Int.AR). Her book *Time Matter[s]: Invention and Re-imagination in Built Conservation. The Unfinished Drawing and Building of St. Peter's in the Vatican* was published by Ashgate in 2013. She is currently editing a book, authored by Marco Frascari and titled *Marco Frascari's Dream House: A Theory of Imagination*, which is forthcoming with Routledge in 2017. She holds a 'Dottore in Architettura' from the University of Genoa, Italy. She is a licensed architect in her native country of Italy.

Nigel Green is a photographer, artist and lecturer and is currently teaching the HND Photography course at Sussex Coast College Hastings. He has exhibited and published photographic projects that document genres of modernist architecture in the United Kingdom and Europe. His monographs include *Reconstruction*, a photographic documentation of the post-war architecture of Picardy in Northern France, published by Diaphane in 2010, and *Dungeness*, which documents the power station complex at Dungeness in Kent, published by Photoworks in 2003. In 2008, he completed a PhD at the University for the Creative Arts, which looked at the relationship between photography and the representation of modernist architecture. Photographic commissions include *Mind into Matter* for the De La Warr Pavilion in 2009 and *From Eros to the Ritz* for the Royal Academy in 2012. He also works in the collaborative art practice Photolanguage, which explores experimental approaches to the documentation of architecture, landscape and urban space.

Tim Ireland is a UK-registered architect, with experience working as a senior and project architect in small-scale private and large-scale international practices. Tim was awarded an EPSRC research grant and left commercial practice to concentrate on his PhD in architecture and computational design at the Bartlett School of Graduate Studies, University College, London. Tim is currently Senior Lecturer at the Leicester School of Architecture where he teaches design and theory and runs the post-graduate Motive Ecologies programme, which amalgamates architecture with biology, semiotics and computational thinking. He also heads a boutique architectural practice in London, where he lives. Tim's research is a synthesis of algorithmic and biological design thinking applied to the conception of architectural space in order to generate architectural scenarios that are driven by and articulate life's artefact-making processes.

Davina Jackson is a Sydney-based writer, editor, curator and project strategist who develops and promotes new transdisciplinary urban development themes through international books, blogs, exhibitions and events. Davina has been a Visiting Research Fellow with Goldsmiths computing at the University of London and a former multidisciplinary design Professor with the University of New South Wales. She was a founder of the world's first 'smart light' festivals in Sydney (Vivid) and Singapore (iLight Marina Bay). Her recent books include *SuperLux: Smart Light Art, Design and Architecture for Cities* (Thames and Hudson, 2015) and *D_City: Digital Earth | Virtual Nations | Data Cities* (a report on the environmental simulations movement, sponsored by the Group on Earth Observations, 2013). She curated two recent exhibitions with the City of Sydney's Customs House (*Spaceship Earth*, 2014 and *SuperLux: Smart Light Cities*, 2015). Her next book is a biographical monograph on Douglas Snelling, an England-born, pan-Pacific modernist architect and designer (Routledge, 2016).

Scott McQuire is Associate Professor and Reader in the School of Culture and Communication at the University of Melbourne, Australia. He is one of the founders of the Research Unit for Public Cultures, which fosters interdisciplinary research at the nexus of digital media, art, urbanism and social theory. Scott is the author of several books including *Visions of Modernity: Representation, Memory, Time and Space in the Age of the Camera* (1998) and *The Media City: Media, Architecture and Urban Space* (2008), which won the 2009 Jane Jacobs Publication Award offered by the Urban Communication Foundation. His new book *Geomedia: Networked Cities and the Future of Public Space* will be published by Polity in 2016.

François Penz is an architect by training and Professor of Architecture and the Moving Image at the University of Cambridge where he directs the Digital Studio for

Research in Design, Visualization and Communication. He is the Director of the Martin Centre for Architectural and Urban Studies – the research arm of the Department of Architecture. He is also a Fellow of Darwin College. He has written widely on issues of cinema, architecture and the city and recently co-edited *Urban Cinematics: Understanding Urban Phenomena Through the Moving Image* (2011). His monograph on *Cinematic Aided Design: An Everyday Approach to Architecture* will be published by Routledge in 2017. He is also co-editing a book on *Cinematic Urban Geographies* to be published by Palgrave MacMillan in 2016. Since 2010, he has been the co-director of the Cambridge – Nanjing Research Centre on Architecture and Urbanism. In the summer of 2015, he produced the film *Recovering the Lost Church of San Pier Maggiore* as part of the National Gallery's exhibition *Visions of Paradise*.

David Ross Scheer received his master of architecture degree from Yale University in 1984. He is currently Associate Research Professor of Architecture at the University of Utah. He has taught architectural design, history and theory at the University of Cincinnati, Arizona State University and Miami University (Ohio). His primary research interest is in the effects of digital technologies, chiefly building information modeling (BIM) and computational design, on design thinking. He was a member of the national advisory group of the American Institute of Architecture's Technology in the Architectural Practice Knowledge Community (TAP) for seven years and was its Chair in 2012. He published a monograph titled *The Death of Drawing: Architecture in the Age of Simulation* (Routledge, 2014). The book explores changes in architectural knowledge brought about by the use of digital technologies to visualize and develop architectural projects – changes which may be summarized as a shift from representation to simulation. In parallel with his research, Mr. Scheer has maintained an active architectural practice for over 30 years. He co-founded the firm of Scheer & Scheer in 1994. The firm's award-winning work includes a variety of building types as well as urban design projects.

Nicholas Temple is Professor of Architecture at the University of Huddersfield. Previously, he was Professor and Head of the School of Architecture at the University of Lincoln and Assistant Professor of Architecture at the University of Pennsylvania. A graduate of Cambridge University, Temple won the Rome Scholarship in Architecture (1986–88) and was a Paul Mellon Rome fellow in 2012. A qualified architect, fellow of the Royal Society of Arts, the Royal Historical Society and an Academician of the Academy of Urbanism, Temple was recently shortlisted for the International CICA Bruno Zevi Book Award 2014 for his publication *renovatio urbis: Architecture, Urbanism and Ceremony in the Rome of Julius II* (Routledge, 2011). He has published extensively on the history and theory of architecture and urbanism, including *Disclosing Horizons: Architecture, Perspective and Redemptive Space* (Routledge, 2006), and co-edited *The Humanities in Architectural Design: A Contemporary and Historical Perspective* (Routledge, 2010), *Architecture and Justice: Judicial Meanings in the Public Realm*(Ashgate, 2013) and *Bishop Robert Grosseteste and Lincoln Cathedral: Tracing Relationships between Medieval Concepts of Order and Built Form* (Ashgate, 2014). He was recently a recipient of a British Academy/Leverhulme small grant award to research the commentaries of Lorenzo Ghiberti.

Introduction

Reproducing reality – a psychological origin of the technological visualisation of space?

Graham Cairns

The book builds on the work I have engaged with for the past twenty years, in which I have investigated and published on architecture as a phenomenon integral to our understanding of twentieth century visual culture. Specifically, the origin of the book's concerns come from this particular approach to examining this architectural-visual culture interplay. It is an approach in which the tropes of the technologies through which we see and represent the world around us, and thus the architecture we conceive and build, is interrogated as a factor informing the nature of human vision at any given time. By extension, these 'technological tropes' are seen as then manifesting themselves – through modifications to sight – in architectural discourse and production. It is an approach I have defined as neo-formalist.

Within the conceptual frameworks laid out by this neo-formalist approach, any consideration of how technologies of sight can be said to have influenced architecture over time runs parallel to the argument that certain patterns of technological motivation and evolution repeat themselves with the emergence of every new technology of sight. In this case, the pattern identified is that visual technologies tend, in their early years of development, to advance from the basis of attempted mimicry. What they tend to mimic is what the eye perceives in all its optical fidelity. Hence perspective sought spatial realism optically, photography reproduced the eye's imagery visually, and film recreated forms and spaces realistically in not only visual but also temporal modalities.

Following this line of argumentation, technologies of sight pass through a developmental phase of 'optical echoing' in which the potential of their technological capacitates visually are subsumed by an obsession with the perfection of imitation. It is only once the visually reproductive challenge has been met and superseded that the artistic and technological potentials of the medium open up to its full potential. At this point, the images they produce are free to explore a new visual and optical terrain, and the possibilities of them shaping other human vision and artistic form emerges.

This understanding of a form of Hegelian dialectic in the advancement of visual technological representation was key to my previous works and in which I expressed, with reference to film, in the 2013 book *The Architecture of the Screen: Essays in Cinematographic Space* thus:

> Film then, is nothing more than the most recent stepping stone in the long evolutionary line of "technological sight". Today, we contemplate the completely digitised visual world appearing on the horizon from the vantage point it offers. In its privileged position of near distance, film is perhaps the most important

precedent we have today for what this fully digital world will bring. Completely understanding this precedent and its influence on architecture, and on society at large, may never be possible; in particular when considered as a cross disciplinary phenomenon, as it has been here. Nevertheless, it is still worth reminding ourselves of the radical potential it was once seen to have.

Key to its early radicalism was its new visual language and its mechanical capacity for "realistic representation". As we have seen in this work however, it was also a medium with its own optical and cinematic vocabulary and an ability to represent the world in motion. It would be these characteristics that would allow it to reconfigure what it captured in its lens. It was this that made it able to present the world in totally new on-screen compositions. Central to film's impact on architecture then, was its optical syntax. This, we suggest, may be much more significant than anything it actually represented on screen; a room, a building, a city and it is this that we emphasised throughout these pages.

This dualistic ability to "recreate reality" on the one hand and "create the incredible and the impossible" on the other, also characterised photography and perspective drawing before it. It most certainly characterises the visioning technologies developed in recent years. As with film, both perspective and photography moved beyond their mere technical ability to "reproduce reality" very quickly. Both mastered perceptual representation and immediately entered the realm of "perceptual creation". In the case of perspective drawing it would manifest itself in the illusionism of the Baroque, whilst in photography it would be seen in the fragmentary and dynamic spatial compositions of the 1920's New Objectivity.

This is perhaps key to understanding the path current developing technologies will follow. Just as painting moved beyond its literal representation of the optical world, when an improved reproductive technology emerged, so too in turn, did photography. In this instance, it was film that played the role of usurper. In each case, the fascination with realism was mastered, absorbed and eventually morphed. It emerged as an interest in the use of visual technologies to "create"; to fabricate what could not be seen or experienced by the naked eye. In this regard, the history of art gives us a clear example of a Hegelian evolutionary process. Reproduction is followed by deliberate distortion.

It may not be a phenomenon restricted to the arts however. If we consider the realm of robotics; the initial goal set by science is the reproduction of the human form and the capacities of the human body. Similarly, artificial intelligence represents a scientific endeavour based, in the first instance, on the mimicry of the human mind and its processing functions. Virtual reality is another example. Here, the "reproductive" aims of the technology in question are directly referenced in its terminology. Reality is to be recreated, only not quite. Taking the metaphor to its extreme, we find in these realistic reproductions the human tendency to play God; for "man" to reproduce "man" in his own image. Where such things will lead once genetics achieves its own particular "reproductive" ends, remains to be seen. Here too however, some see the same characteristics in play.

Emerging from an examination of the filmic medium, the evolutionary dialectic of optical mimicry ceding to the emergence of a new visual language expressed here was, from the outset of this volume, a key idea in my development of this book. Indeed, in setting out the terrain of this volume to individual contributors as one upon which we explore and document the multifarious ways in which technologies of sight have informed architectural thought, conception and representation through time, I was explicit. It was put to the author of each essay in the following terms:

> The premise of the book is that 'visioning technologies' have tended, in their incipient moments, to repeat one aim – the reproduction of reality. Perspective froze space visually, photography captured it momentarily, film presented it in time and virtual reality immerses us in it holistically. Even parametricism can be said to reproduce a 'reality' on screen – it allows us to watch the real-time process of form formation (what we previously called design).
>
> However, more than just reproducing reality, these technologies influence architectural design, theory, and intellectual/spatial conceptualisations in a way that evolves over time. In the case of perspective drawing, the influence of the 'new mechanical drawing technique' would manifest itself in the single point perspective images of Brunelleschi, feed into the focal point perspective spatial compositions of the Renaissance, and evoke a concomitant reconsideration of our place in the world. Eventually, it would be used to transform spatial perception specifically through the illusionism of the Baroque and related notions of advanced humanism.
>
> In the context of photography, the reproductive potential of the image was, for Reyner Banham, what made the International Style, *international*. However, by the end of the 1920s, the angled imagery of New Objectivity Photography was deliberately *over*-emphasising the compositional dynamism of the early Modern Movement – an approach that can be argued led to the promotion of an ever-more spatially complex *modern* architecture and a mechanised conceptualisation of the contemporary human psyche.
>
> In both cases, we have technologies that in their first iterations 'reproduced what the eye could see' – it was this ability that made them 'revolutionary' in their day. However, both technologies quickly moved beyond their mere technical ability to 'reproduce reality' – they both mastered their own forms of perceptual representation and immediately entered the realm of 'perceptual and architectural creation'. In doing so, they often recalibrated standard intellectual understandings of both space and 'the human'. This book is initiated on the belief that we may be able to trace out how this dynamic has repeated itself with the emergence of almost every new 'technology of sight' and how it may be repeating itself today in the age of digital imaging.
>
> By documenting the historical influence of 'technologies of sight' and applying its template of analysis to contemporary technologies – and the design tropes that stem out of them – this books hopes to participate in the 'construction' of a history for a current generation of architects. This generation of architects is 'reproducing and visualising realities' through digital visualisations, virtual reality environments, and the real-time digital formation of parametricism. This book responds to the fact that they are embracing the radical potential of the latest visualising techniques of the digital age without a fully explored historical background within which to see their work. In identifying the outlines of this history, this book will trace out a thread of architectural theory that has yet to be fully explored and exposed, but which is of direct contemporary relevance.

In building on this provocation as their starting point, the contributors to this book have taken their own particular interpretative line of analysis. It has led them to consider a multifarious array of social, representative, and productive consequences of their own particular technology of study, its acceptance, and subsequent application in the architectural context. For some, the issue at play is precisely this tendency for technologies to mimic optical reality that is of importance, whilst for others, it is their introduction of new visual tropes to the architectural lexicon. By contrast, some authors have developed this provocation into a consideration of the direct formal influence the visual language of their particular technology of study has had on architectural form. Other authors look at these formal effects from a broader social perspective, considering the public perception of architecture resulting from the advanced influence of technologies. All, in one way or another, identify that the representation, conception, design, or perception of architecture has been altered by different visual technologies over time.

Thus *Visioning Technologies – The Architectures of Sight* is a collection of texts from theorists and practitioners that examine how architecture has been, and is, reframed and restructured by the visual and theoretical frameworks introduced by different 'technologies of sight'. It sets out four sections that respond to approximate historical periods but, more specifically, four visioning technologies: perspective, photography, film, and digital media. In each section, authors deal with their own area and historical period of expertise, with the intention of, together, marking out and analysing the historical and contemporary territories in which architecture has been transformed by technologically induced shifts in human perception from the fifteenth century until today.

In commissioning and placing the work of these experts in a historical timeline, albeit a loose one, I am attempting to instigate with this book a formalising and framing of our understanding of the varied ways in which 'technologies of sight' have influenced architecture over time. It is a particularly important moment to do this, as in the current age of digital visualisation and architectural production, the technologies being employed are often defined as 'new'. What this book attempts to do is suggest that certain underlying trends, motivations, and patterns of thought and production are actually simply being repeated in this 'new' age. In doing this, the book clearly risks, indeed embraces, the possibility of being defined as an 'operative history'.

Both the historical categorisations as well as its technological classifications used to guide the book's structure can be seen as reductive – erasing the inevitable cross-overs and nuances that exist in the history of intellectual and practical evolution of any discipline or line of enquiry. Accepting this as almost inevitable, the book's structure is seen as a useful, if not totally necessary, framework that facilitates an understanding of the history of architecture's relationship with technologies of vision. A book resulting from these heuristic categorisations has its limitations and may be more safely defined as a first step in our presentation of this history than a definitive history of architecture's visioning technologies. However, it has its utility which, one hopes, will be acknowledged.

The history hinted at here, then, has served as the calibrating device for the inclusion of authors from different disciplines in this volume. They are theorists and practitioners whose work has been brought together in four main sections. The first section deals with the drawing techniques of the fifteenth to eighteenth centuries with particular emphasis on perspective drawing as a graphic technique premised on what the editor defines as "the visual reproduction of reality". Opening this section is Nicholas Temple with an essay entitled 'Envisioning Geometry: Architecture in the Grip of Perspective'.

Framing his essay in what he calls "the saturation of sophisticated imagery in our digital age", he expressly discusses the optical closing of the gap between reality and illusion contemporary technologies allow as a phenomenon that "lulls the senses about what it means to represent something in the contemporary world". For Temple, there is an important role for memory (history) today which he seeks to underline by returning to, and building upon, the work of Dalibor Vesely, particularly by examining the epistemological and ontological relationships between pictorial representation and the late medieval tradition of 'perspectivist optics' – a relationship that had notable consequences for our understanding of the representation of architecture.

Following on from Temple is Nader El-Bizri, whose text 'Desargues' Oeuvres: On Perspective, Optics and Conics' examines the visioning techniques in the perspectival and projective geometric methods of Girard Desargues. Underlining Desargues' aims in supporting the "construction of architectural space through the mathematical underpinning of its visual representation with greater precision and objectivity", El-Bizri presents a clear historical precedent for how the visualising techniques we employ feed into the architecture we build. Suggesting that key to this was not only the formal qualities of Desargues' imagery, but their acceptance and dissemination through social channels, he also gives a sense of how techniques of representation do not engage with architecture exclusively on formal terms.

In 'Galileo's Limit: Mechanical Sciences' Technologies of Sight and the Translation of Analogical Representations into Diagrammatic Illustrations', Federica Goffi offers what she calls "a micro-historical close up" of sciography, axonometric drawings, and vertical sections in the works of Leonardo da Vinci, Galileo Galilei, and Giovanni Alfonso Borelli. Suggesting that this reveals "a telling transformation from analogical representations to diagrammatic illustrations", she sheds light on the role of drawing techniques in the development of theories of structures. Linking this to a shift from anthropomorphic architecture to an emphasis on mechanical engineering in this time period, she explores the visual alongside the productive and the physiological.

Revealing a range of ways in which drawing techniques, and particularly perspective, informed architecture directly and nuanced that informing process as they themselves evolved in response to conceptual and technical advancements, these authors set the origin of the book's timeline. They also lay out the most obvious historical foundations upon which it seeks to support itself. Leading us up to the eighteenth century, they cede way to what can be considered the next major development in visual representation technologies: the photographic image in the nineteenth century. Beginning this particular section of the book is Nigel Green, with a text entitled 'The Transformative Interface: Fragmentation, Process and Construction in the Photographic Representation of Architecture'.

Picking up the history of photography and its relationship with architecture in the early years of the twentieth century, Green considers the medium at a point in time when its ability to 'reproduce reality' was well assimilated. Concerned with a more evolved point in the development of the photographic image, it is the capacity of the still image to construct new visual realities that is of interest to Green. Locating his test in the key modernist discourses surrounding photography in the 1920s, Green identifies four categories of photographic practice, each of which reveal a different facet of photography's contribution to architectural thought and forms in this period: the spectacular, the experimental, the documentary, and the hybrid.

In '"The Great Publicist of Modern Building": Photography and Architecture in the Inter-war Years', Valeria Carullo, director of the RIBA photography library, takes a more historiographical approach to the same time period. Citing concerns as early as 1932 that the architectural professions use of photography opened the door to the possibility that "young architects might try to design buildings that look to the human eye as the photographs", she explicitly references historical debates about the impact photography on architectural design itself. Returning to inter-war editions of *Architectural Review* and the New Photography tropes and terminology of extreme angles, abstraction, and 'the unfamiliar', she constructs a reading of architecture's assimilation of the photographic image that is both technical and social.

Also referencing historical architectural photography, but using it to inform our readings of contemporary photographic production and influence in the context of architecture is Iñaki Bergera. Bergera's contribution, 'Photography and Architecture: From Technical Vision to Art and Phenomenological (Re)vision', commences its considerations in the "context of the oversupply of architectural imagery today". In this context, his reading is as much social as it is technical, suggesting that the technological reproducibility of the digital image, and the ease of its electronic dissemination through the Internet, results not only in the "aesthetic manipulation of the image" but also the "anesthetisation of architects, users and public opinion". Suggesting, then, that the various technologies of photographic production and distribution should be considered in tandem, he gives an outline of the "contemporary theoretical, cultural and artistic context in which architectural photography operates", arguing on the one hand that a greater 'realism of use' should be evident in architectural photography and, on the other, that art practice may contain ideas of use to architectural photographers who wish to use the medium more innovatively and constructively in this saturated social context.

In the third section of this book, we move onto a technology of vision born at the very end of the nineteenth century and whose influence on architecture and artistic and popular culture more generally would be fundamental throughout the twentieth century. The first engagement with it in this book comes from François Penz who in 'Absorbing Cinematic Modernism: From the Villa Savoye to the Villa Arpel' focuses on the historical engagement of modern architecture and film through the work of Le Corbusier and Jacques Tati. Considering both the historic and contemporary implications of this engagement, Penz offers a reading of modernism and our contemporary understanding of it that is inseparable from the filmic medium. He suggests that Pierre Chenal's film *Architectures d'Aujourd'hui* (1931), and in particular the scene of Le Corbusier's La Villa Savoye, is pivotal to gaining an understanding of a history of architectural modernism through cinema. However, he also suggests that Tati's interpretation and, importantly, cinematic presentation of modernism has become embedded in the architectural psyche of the modern, as manifest and analysed in the twentieth and twenty-first centuries, respectively.

Highlighting the idea that the manner in which film engages with, and directly informs, architectural thought and practice is an issue of relevance and potential today, Graham Cairns contributes a chapter entitled 'Fragmented Fluidity: A Possible Future for Spatial Theory and Praxis in Filmic Form'. Suggesting that the reawakening of the filmic-architectural analogy and interaction instigated by Bernard Tschumi in the 1980s around the work of Sergei Eisenstein was premised on a lack of appropriate modern filmic spatial experiments, he calls attention to a series of films that emerged between 1996 and 2006. With particular focus on *Run Lola Run*, 1996, he argues that these films offered potential precedents for the architecture of deconstruction, as proffered by Tschumi, and that they could be seen as useful formal tools and precedents for an even more contemporary set of architects, the parametricist movement, which today celebrates folding and morphing architectural forms and form production whose aesthetics, spatial, and conceptual templates are all echoed in a series of filmic experiments that emerged in the period in question.

Finally, returning to a referencing of the theories and projects of the early twentieth century period with which this section began, albeit with a more explicit emphasis on László Moholy-Nagy, Scott McQuire uses this historical context to build quite a different argument to that laid out by Penz. In 'Intersecting Frames: Film + Architecture', McQuire suggests that the conceptual possibility of film merging with architecture discernible in this period has recently been reconsidered as "digitizing photographic and video images [have become] commonplace." In this context, McQuire seeks to build a reading of film as "the orchestration of mobile and dynamic fields of vision" and, as such, suggests it "remains not only useful but essential to understanding the new conditions . . . that continues to influence contemporary architecture." Moving beyond the purely technical characteristics of the medium, or those of contemporary technologies that have to an extent superseded it, McQuire's reading of film leads to speculations about digital architecture and the contemporary technologically laced, 'smart city'.

McQuire's commentary on film and architecture in the digital age is a perfect segue to the final section of this book in which digital technologies are examined in detail. David Ross Scheer offers 'Hyperreality, Vision and Architecture', a text which he frames in an "age under the hegemony of a mode of visuality in which the ontological distinction between images and experience has been dramatically blurred, if not entirely eliminated." Arguing against the use of digital technologies to present – and to an extent design – buildings in all their 'hyper-real detail' through computer renderings and animations, he proposes that the technologies facilitating this are "emptying all experience of reference". Furthermore, our new technological capacities are making our knowledge purely "operational", reflecting "Western society's tendency to replace meaning with performance". For architectural design, this means a shallowing of our engagement with the spaces we conceive as architects and designers. For Sheer, the potential influence of the technologies of simulation on architecture are problematic, and he argues for their containment through more varied modes of 'representation'.

Thinking in a very different register is Davina Jackson. In her text, 'Rebooting Spaceship Earth: Astrospatial Visions for Architecture and Urban Design', she examines cutting-edge science in the context of Digital Earth and Earth observations that form part of the Observation System of Systems (GEOSS) intergovernmental scientific programme launched in 2002. In that context, she suggests that the advances offered by science to "a sophisticated global system of simulating environments are not yet effective for architects". She also indicates that the contexts in which simulation in architecture does occur is far from ambitious. Even at its most advanced, in the computer-generated imagery studios of companies such as Pixar and Weta, the ambition is limited – in this case to "visualising fantasy environments and manipulating imagery of real cities". What her chapter argues is that the technological potential of new technologies being developed at the cutting edge of science will go beyond the 're-creation' of fantasised existing realities and give architects and urban designers a whole range of new options for both research and design – possibilities that will be far more significant than even a new visual language.

This argument that the technologies of vision we have at our disposal today and which are emerging at a seemingly ever faster pace will eventually do more than simply alter our visual, formal, aesthetic – and even social – frames of reference is repeated in the final essay of the book by Tim Ireland. In 'Leveraging Nature to Envision (Functional) Space: An Architecture of Machinic Abduction', Ireland, more than any of the other authors collated here, suggests that the digital technologies available today potentially stretch our "obsession with sight" to realms that leave behind our concern with visual languages, composition characteristics of form and aesthetic architectural concerns. Setting his arguments in the context of contemporary 'visioning technologies' such as virtual reality and hyper-real visualisation, he proposes that the scientific knowledge behind these technologies and approaches will not only master our ability to visually reproduce architectural forms, whether existent or imagined, but also begin to mimic, explore, and exploit non-visual realities. In his case nature itself.

The final chapters of this book suggest that the technologies at our disposal for the visualisation, design, and analysis of architecture are evolving. These 'newer' technologies are, through hyper-real visualisations and virtual realities, repeating the tendency described in the early statements made in this book and its provocation to its authors. However, they are seen as capable of much more. They are seen as capable of mimicking more than the visual form of the spaces we seek to design and inhabit. As a result, the implications they have go beyond the development of a new visual vocabulary and its social dissemination and appropriation.

In this conceptualisation of our current situation, contemporary digital technologies are capable of altering the course of architectural thought and production in ways not conceivable to previous technologies of sight which, because of their own particular limitations, were more limited to the 'purely visual'. Whether these hypothesises prove to be true and whether the visual culture many of our technological developments over two millennia have thus far been based upon and formed are actually usurped is far from clear at this historical juncture. However, what is clear from the texts collated here is that our technologies of vision thus far have repeatedly overlaid a new optical lexicon on the established visual vocabulary of the world we inhabit. The fact that this informs the way we see spaces, the way we look at buildings, the way we represent cities, or the way we design and create the places we inhabit is not surprising. Our technologies of vision inevitably form both our architecture and our sight.

Neo-formalism: The technological and social evolution of visual representation and the emergence of contemporary optics and architecture

Graham Cairns

Neo-formalism and the architectural analysis of film

The argument that our technologies of vision inform both our architecture and our modes of vision, or, more precisely, our ways of looking, has been at the heart of my work in architecture and visual culture for two decades. It was central to the emergence in my thoughts and writings of the notion of *neo-formalism*. This possibly retrograde term refers to a methodology of analysis I initially developed through a series of works on film, both video installations with Hybrid Artworks in the 1990s and a range of theoretical essays elaborated over a number of years and terrains. The objective of formulating this methodology was to aid in identifying and explaining the influence of twentieth-century visual culture and its technologies in the evolution of modern architectural theory and forms. In the course of over fifteen years of research in this field, it is an approach that has been complemented by an analytical framework drawn from the social art history tradition.

The initiation of the idea of neo-formalism as a definable form of analysis first appeared after the publication of my 2007 book on film and architecture, *El arquitecto detrás de la cámara: una visión espacial del cine* (The Architect Behind the Camera – A Spatial Vision of Film) in Madrid, Spain. It was adapted to the context of the promotional photograph and its relation to commercial architectural development of the twentieth century in my 2010 book *Deciphering Advertising, Art and Architecture: Selling to the Sophisticated Consumer*. More recently, I have reworked these ideas in *The Architecture of the Screen: Essays in Cinematographic Space*, 2013. In each of these volumes, the underlying concern with examining and explaining the emergence of visual tropes through 'technologies of vision' and their influence on architecture is manifest.

Loosely, neo-formalism harkens back to the Wölfflinian tradition and an underlying parallel concern with the evolution of artistic style and human vision. More specifically, it builds on this tradition as applied in the filmic context by Rudolf Arnheim, and, principally, his understanding of the moving image as an 'artistic' form requiring viewing subjects to apply a specific set of perceptual modalities, *Film as Art*, 1969. However, it also encompasses ideas of post-formalism, as outlined by Whitney Davis in his essay 'What Is Post-Formalism?' to underline the role of art forms and objects themselves in the evolution of the artist's, designer's, or architect's own particular 'way of seeing'. It is seen, then, as an analytical framework that facilitates our understanding of how modern visual medias and their perceptual characteristics can, and have, influenced the evolution of the contemporary vision and thus contemporary architectural thought and, finally, form.

Emerging from an initial analysis of film and its influence on architecture in the early twentieth century, the neo-formalist system of analysis I outline here results in an interpretation of the birth of film, as with other technologies of sight before it, as the nascence of a new visual language. Its application facilitated attempts to understand how film's 'visual language' reframed space on screen and through that how the cinematic medium's new spatial conceptions and visual vocabulary influenced architectural theory and practice in the twentieth century.

In this regard, it drew on the formalist tradition in film studies, specifically the writings of Rudolf Arnheim (*Film Art*, 1969, *Visual Thinking*, 1969, *Art and Visual Perception*, 1974), Hugo Munsterberg (*The Photoplay: A Psychological Study*, 1969) and Béla Balázs (*Theory of the Film: Character and Growth of a New Art*, 1949). Aspects of the analytical templates bequeathed by these intellectual giants were used to inform the understanding I was tracing out of the ways in which representations of 'movement', and its perception, could directly and indirectly predispose architects to conceive and design in particular ways. Forming examples of this, the constructivist architecture of the Soviet avant-garde's celebration of motion, or Le Corbusier's cinematic reading of the 'promenade' – a spatial and temporal device Beatriz Colomina defines in *Privacy and Publicity: Modern Architecture as Mass Media*, Cambridge, 1994 – resulting from modern eyes, i.e. 'eyes that move'.

In reaching back to an examination of film through the formalist analysis of Arnheim, Munsterberg and Balázs on the one hand and the architecture of the early twentieth century avant-garde on the other, the early application the neo-formalist approach schematically outlined here interprets film as a medium 'freed of the photographic image from its state of stasis'. In doing so, it permits us to argue that the medium of film introduced a new set of visual tropes into the lexicon on the modern world at the turn of the twentieth century and that this expanded the visual vocabulary of artists, feeding into the architectural mindset of the day. Early film, I suggest, led to a questioning of how we perceive the space around us; it offered new possibilities in our understanding and representation of buildings and presented architects with a platform for optical and spatial experimentation at the scale of the interior, the building, and the city. In short, it was a visual language that influenced the evolution of architectural form.

The neo-formalist modality of thought, then, is inspired by an architectural engagement with the discipline of film from the perspective of its early formalist theories. It is inspired by a period that gave rise to theories which considered the practice of the cinema as analogous to that of architecture – a period in which Dziga Vertoz would call himself Kino Eye and defined it as "an eye that constructs". To explicitly examine how film's visual language can inform the architect's visual vocabulary according to this neo-formalist template, I carried out a series of spatial and cinematic dissections of film scenes in workshops and studies between 2002 and 2005. Documented in *The Architecture of the Screen*, 2007, these led to the definition of a concept referred to as 'cinematographic space' – an understanding of spatial perception created on screen by film directors – and how this differs from, and can inform, the spatial experience of the 'physical'.

Premised on this analysis of 'cinematographic space', and defining the analytical processes and speculations it encompasses as neo-formalism, I have applied the readings of film and architecture it elicits to the kinetic concerns of architects such as Alexander Vesnin and Konstantin Melnikov in the 1920s, the architectural promenade as a movement sequence in the work of Le Corbusier, the interest in sequential experience in the designs of Carlo Scarpa, the arguments on architectural representation found in the writings of Robert Venturi from the 1970s, and the optical effects created in the architecture of Jean Nouvel in the 1990s, amongst others. Using this approach, we are able to shed new light on the history and existence of a particular sub-strand of modern architectural theory and practice in which the implicit influence of film's visual lexicon on the discipline of architecture is made manifest. It is not, however, limited to formal analyses and subsequent transpositions to architectural practice in the exclusive context of the cinematic.

The formal and social construction of spatial readings

Despite its origins in attempts to better understand and examine the influence of film on architectural evolution, the aligned set of premises and systems of analysis being defined here as neo-formalist were also applied in examinations I carried out on a quite different phenomenon in the early years of this millennium: the consideration of visual advertising and its implications for the models of commercial architecture that have evolved throughout the second half of the twentieth century. In the book *Deciphering Advertising, Art and Architecture: Selling to the Sophisticated Consumer*, 2010, I attempted to demonstrate that the neo-formalist concern with the mutual evolution of visual tropes, human vision, and architectural form was applied alongside a more explicit understanding of how social factors affect this relationship.

Applied to an analysis of the still image – more precisely the commercial photograph of the advertising image – the arguments that emerged from it in this book involved an interpretation of the visual advertising of the twentieth century as having passed through three distinct stages of visual evolution, each of which overlaid one system of visual reading upon another: an evolution that engaged semiotics, post-structural theory, and, more recently, phenomenology as systems of visual organisation and communication through the photographic image. When analysed as perceptual modalities aligned with specific visual tropes employed in still imagery, a neo-formalist reading of these images identifies how their optical and perceptual tropes could be, and were, transposed to architectural form.

The visual analysis offered in this particular neo-formalist approach thus moves beyond the purely formal to draw on the classic semiotic texts Roland Barthes and Judith Williamson, as well as more recent authors from within this canon of visual interpretation such as Erik Du Plessis' *The Branded Mind*, 2011. From the post structural field, it adapted the writings on advertising by Goldman and Papson (*Sign Wars: The Cluttered Landscape of Advertising*, 1995) by aligning them with the theories of Umberto Eco

(*A Theory of Semiotics*, 1973, and the *Role of the Reader*, 1979). From the field of phenomenology, the functioning of the human perceptual apparatus, as explained by Maurice Merleau-Ponty in The *Phenomenology of Perception*, 1960, was applied in the commercial context of 'hidden advertising', as defined and discussed by Edwin Baker (*Advertising and a Democratic Press*, 1993) and David Michie (*Invisible Persuaders*, 1998). This series of theoretical pairings, I argued, was key to understanding the evolution of the advertising strategies employed, their manifestation as visual tropes in photographic imagery, and, ultimately, their manifestation in architecture through the selective commissioning of architects able to work within the semiotic, post-structural, and phenomenological visual frameworks set up by the advertising industry across this period.

The architectural context of this study was the most explicit realm for the manifestation of these visual communicative devices: the retail store. Thus I examined the aesthetic and formal attributes of the commercial work of Norman Foster and David Chipperfield in the early 1980s through the prism of semiotic systems of visual communication, the late 1980s and early 1990s projects of Nigel Coates and Jun Aoki were analysed in tandem with the 'post-structural promotional practices' of deliberate visual dissonance in the photographic imagery of the period, and the 'iconic' early twenty-first century commercial/retail works of Rem Koolhaas and Frank Gehry were considered in the light of the phenomenological theories (re)adapted by advertisers in the past two to three decades to elicit what Merleau-Ponty described as 'distracted attention'.

Distinct from the more formally focused analysis of film offered in my earlier writings, it was not exclusively the compositional qualities of the photographic image that were analysed here but also our modalities of reading images through different visual communitive systems. These images were thus read as optical-social constructs that both rely on the eye functioning according to given stimuli and the brain interpreting stimuli according to social conventions. Their subsequent influence on architectural form was similarly dualistic.

Commencing its investigation with a background analysis of images whose optical-social construction was defined as semiotic, this neo-formalist study reached back to the 1960s – an epoch in which commercial architecture was reduced to a direct imitation of the forms and aesthetics of visual advertising and, thus, almost exclusively, considered as a one-dimensional visual messaging system. This one-dimensional form of architectural image, I argued, dominated the commercial sector until the 1980s when semiotic visual tropes in the photographic image were overlaid by a new range of advertising strategies emerging during that period: the post-structuralist *opaque ad, non-ad,* and *self-referential ad.*

With the emergence of new visual techniques in commercial photography, a concomitant shift became discernible in parallel analyses of architecture. Considered and defined as 'post-structural' because of their deliberate multi-narrative approach to optical-social communication, I argue that the advertising imagery of this period informed the commissioning of designers such as Nigel Coates in the United Kingdom, whose own brand of multi-narrative architecture could be adapted as an ideal architectural foil to this wave of visual tropes and advertising theory.

In continuing this formal and social analysis of advertising imagery and its influence on commercial architecture in the period 2000–2010, attention was focused on the (re)emergence of 'hidden advertising' and its influence on the commissioning of the commercial 'hybrid' in this period. In this light, the stores of Herzog & de Meuron, John Pawson, and OMA, for example, are formally dissected as spatial and aesthetic constructs but equally analysed through the phenomenological concepts developed by Maurice Merleau-Ponty, in particular, the notion of selective attention. Overlaying a formal analysis with a phenomenological one, I define the influence of advertising's visual tropes on the commercial architecture of the late twentieth century d in terms of a significant paradigm shift in the commercial architectural context.

Social origins in the technological architecture of vision

The works on film and advertising referenced here involve the consideration of a very different set of examples to those that inform my writings on film. However, both approaches to analysing architecture in the context of visual culture share a concern for understanding how the way we look, and the media we look through, influence what we see and interpret and thus how and what we build as architecture. In the case of film, the primary focus of the analysis discussed was the visual tropes of the cinematic medium itself – the visual language created by the new technology of vision – and how that fed into architectural production and theory. In the latter, this concern with the visual tropes of technological advancement was overlaid with an analysis of how these tropes lend themselves to controlling the eye so as to facilitate pre-established social readings of both images and architectural spaces.

In the first case, the technology of filmic vision is foregrounded as an influence on architecture. In the latter, what gets foregrounded is the commercial manipulation of the 'technology' of the photographic image – and a physiological understanding of how these images are read and deciphered. This in no way distinguishes one technology from another in terms of how it relates to architectural production. The architectural application of visual tropes from film is not exclusive to the cinematic medium, nor is the social use of photography in the advertising context inherent to the technology of the photographic camera. If we consider film in a slightly different context to its incipient years of formal experimentation in, say, any of the innumerable Hollywood films premised on selling the American Dream, the medium's influence on architecture could be defined as more akin to that expressed here by advertising based primarily on social readings that overlay its particular formal tropes.

To illustrate this point, it is worth considering how the power of film's new visual language to influence architecture in formal ways was only momentarily grasped by architects – periodic re-engagements such as those of Jean Nouvel notwithstanding. The radical visual or optical potential of film identified by people such as Dziga Vertoz characterised the short period of the cinematic medium's incipient years – when its forms remained unfamiliar, its techniques were still 'shocking' and its effects still surprising. However, once consolidated in the contemporary psyche, the formal experimentation of the medium itself gave way to convention quite quickly. By the end of the 1930s, the continuity system of filming had been standardised and exported across the globe with the consequence that the once radical language of film very soon simply supported the telling of stories, which, in the hands of both great and mediocre propagandists alike, sold dreamworlds of opportunity and individual triumphs over adversity in a free new world captured in commercial film.

Barely noticed, ignored, and imperceptible behind the power of aesthetic and narrative interests, the formal potential of film from the 1930s onwards was largely reduced to a social one. From that point on, it was no longer a radical melting pot for spatial theories, but rather a site of spectacular sets or, alternatively, the mundane setting for a narrative focus on action. In the field of architecture, film had largely lost its power to influence our conceptual understanding of space by this period. So too with photography, which, by the time it was first applied to advertising imagery, had already ceased to offer significant formal motivation for the public or architects. By the time it was applied in the studies referenced here, it had been fully subsumed onto our communal optical and cerebral parlance.

In bringing up these issues in relation to photography and film, I broach a return to the pre-narrative period of the medium when the potential of its visual language had not been extinguished. However, in doing so, I do not and cannot divorce this from more social readings of film's influence on architecture, whether they be through the presentation of given architectural works or cities, or a consideration of their role as backdrop to action. The same, too, can be said, in nuanced forms, of perspective photography or newer digital technologies. In considering any of these technologies of sight, we must variously address the tropes of their medium as visual and formal influences on architectural production, but also consider the social contexts and readings of their application to, and effects on, architectural discourse, visualisation, and form production.

We can discuss, and indeed it is important to discuss, the drawing techniques of the fifteenth to eighteenth centuries as primarily concerned with the visual tropes of the representational technique rather than narrative elements to be found in say, perspective painting. However, we should do so acknowledging that, on many occasions, the influence of the drawing techniques they describe lay in their cultural acceptance and practical dissemination to society at large. Similarly, when examining digital technologies in order to highlight the potential influences on architecture of the technologies examined, as per a post-political parametricist position, we should be unwilling to fully divorce that from questions such as the social consumption and interpretation of the architectural images resulting from advancements in the technologies we describe. It is out of this dual frame of reference that I suggest that the technological and social evolution of visual representation and the emergence of contemporary optics and architecture can be most fruitfully considered within a particular frame of reference – one that commences with the formal visual tropes of a particular technology and which follows its analysis through to the ways of seeing it induces and the effects it has on architectural experiences and production, but which never loses sight of the social context in which these processes operate. In short, a neo-formalist approach.

Bibliography

Adorno, Thedor. Ed. Gretel Adorno and Rolf Tiedemann. Trans. Robert Hullot-Kentor. *Aesthetic Theory*. Minneapolis: University of Minnesota Press, 1997.

Appleyard, Donald, Kevin Lynch and John Myer. *The View from the Road*. Cambridge, MA: MIT Press, 1964.

Arnheim, Rudolph. *Art and Visual Perception: A Psychology of the Creative Eye*. Berkeley: University of California Press, 1974.

———. *Film as Art*. Berkeley: University of California Press, 1969.

———. *Visual Thinking*. Berkeley: University of California Press, 1969.

Baker, Edwin. *Advertising and a Democratic Press*. Princeton, NJ: Princeton University Press, 1993.

Balázs, Bela. *Theory of the Film: Character and Growth of a New Art*. New York: Dover Publications, 1970.

Cairns, Graham. *The Architecture of the Screen: Essays in Cinematographic Space*. Bristol: Intellect Books, 2013.

————. *Deciphering Advertising, Art and Architecture: New Persuasion Techniques for Sophisticated Consumers*. Faringdon, Oxfordshire: Libri, 2010.

————. *El Arquitecto Detrás de la Cámara: Una Visión Espacial del Cine (The Architect Behind the Camera – A Spatial Vision of Film)*. Madrid: Abada Editors, 2007.

Colomina, Beatriz. *Privacy and Publicity: Modern Architecture as Mass Media*. Cambridge, MA: MIT Press, 1994.

Davis, Whitney. "What Is Post-Formalism? (Or, Das Sehen an sich hat seine Kunstgeschichte)" *Nonsite.org Issue 7.* http://nonsite.org/article/what-is-post-formalism-or-das-sehen-an-sich-hat-seine-kunstgeschichte

Diller, E., R. Scofidio, G. Teyssot and Diller + Scofidio. *Flesh: Architectural Probes*. New York: Princeton Architectural Press, 1994.

Du Plessis, Erik. *The Branded Mind: What Neuroscience Really Tells Us about the Puzzle of the Brain and the Brand*. London: Kogan Page, 2011.

Eco, Umberto. *The Role of the Reader: Explorations in the Semiotics of Texts*. Bloomington: Indiana University Press, 1994.

————. *Theory of Semiotics*. Bloomington: Indiana University Press, 1976.

Goldman, Robert and Steven Papson. *Sign Wars: The Cluttered Landscape of Advertising*. New York: Guilford Press, 1995.

Hauser, Arnold. *The Sociology of Art*. Chicago: University of Chicago Press, 1982.

Merleau-Ponty, Maurice. *Phenomenology of Perception*. London: Routledge & K. Paul, 1962.

Michie, David. *Invisible Persuaders: How Britain's Spin Doctors Manipulate the Media*. London: Bantam Press, 1998.

Münsterberg, Hugo. *The Photoplay: A Psychological Study*. New York: D. Appleton and Co., 1916.

Petrić, Vlada. *Constructivism in Film: The Man with the Movie Camera: A Cinematic Analysis*. Cambridge: Cambridge University Press, 1987.

Van Leeuwen, Theo, Carey Jewitt and Rumiko Oyama. "Visual Meaning: A Social Semiotic Approach." In *A Handbook of Visual Analysis*, edited by Carey Jewitt and Rumiko Oyama, 134–156. London: Sage Publications Ltd., 2001.

Virilio, Paul. *Open Sky*. London: Verso, 1997.

————. *The Vision Machine*. Bloomington: Indiana University Press, 1994.

————. *War and Cinema: The Logistics of Perception*. London: Verso, 1989.

Wolfflin, Heinrich. *Gedanken zur Kunstgeschichte: Gedrucktes und Ungedrucktes*. Basel: Schwabe, 1947.

Part I
Perspective

1 Envisioning geometry
Architecture in the grip of perspective*

Nicholas Temple

Introduction

In the midst of the saturation of sophisticated imagery in our digital age, where it is often impossible to distinguish between reality and illusion, we remain ignorant of the origins of this uniquely modern phenomenon. This lack of awareness continues to lull the senses about what it means to represent something in the contemporary world and the importance that memory (history) plays in imagining and inhabiting other worlds. Such a situation is perpetuated by Hubert Damisch's observation, "The problem of distance – distance between the point of view and the object perceived, and the distance between the eye and the picture plane, which are two different things – is to all appearances the heart of the question."[1]

Damisch addresses this question to Brunelleschi's famous perspective experiments (of which more will be said later), but it could just as well be applied to the broader issue of the relationship between illusion and reality. To examine the historical and cultural contexts of this phenomenon is to uncover something crucial and mysterious about the advent of the modern world. Dalibor Vesely provides some historical context to this transformation:

> The process of uncovering those foundations leads inevitably into the depth of time, back to the generation of Leon Battista Alberti and Nicholas of Cusa and the formulation of Renaissance perspective, the first plausible anticipation of modernity. By examining Renaissance perspective against the background of the medieval philosophy of light, we can come to understand the ontology of architectural space, which is formed by light before it is structured geometrically.[2]

Two issues emerge from Vesely's statement that have a bearing on this investigation: first, perspective is "the first plausible anticipation of modernity" and second, "the ontology of architectural space" was "formed by light before it is structured geometrically". Both assertions are related since the beginnings of modernity were characterised by a shift in the understanding of space as a setting revealed through 'divine' transcendent illumination to one construed in immanent terms as almost exclusively a problem of mathematics.

In this chapter, I return to Vesely's arguments by examining the epistemological and ontological relationships between pictorial representation and the late medieval tradition of 'perspectivist optics' and the consequences of the transition (from one to the

*This chapter is dedicated to the memory of Dalibor Vesely (1934–2015), teacher, mentor and friend.

other) in our understanding of the representation of architecture. Through a study of the ideas of Leon Battista Alberti, Nicholas Cusanus and Filippo Brunelleschi in early fifteenth-century Florence, I consider how this development was expressed in humanist and theological tracts and rendered spatially. Whilst a significant body of scholarship has been written on the subject of linear perspective, in this investigation, I follow a different line of enquiry by considering how the moment when perspective was first realised signalled in historical terms a decisive shift from the earlier medieval world view and at the same time drew upon this tradition to preserve a still pervasive onto-theological outlook.[3]

Medieval optics and the resurgence of appearance

In Dallas Denery's book *Seeing and Being Seen in the Late Medieval World: Optics, Theology and Religious Life*, the author states,

> Theological debates, more often than not, began with ontological questions connected with the principle of singularity, moral problems involving human free will, and theological problems arising from the nature of beatific vision and God's omnipotence. In addressing these sorts of issues, theologians exploited perspectivist ideas and resorted to visual analogies.[4]

In essence, "knowing something is somehow analogous to seeing something."[5] This claim finds expression in the ideas, for example, of Robert Grosseteste, eminent medieval theologian/scientist and bishop of Lincoln Cathedral (1235–1253). His dedication to pastoral work and to church reform drew analogies from his luminary cosmology and his theories of vision. We find evidence of this in his *Hexaëmeron*, a commentary on the creation narrative in *Genesis*, in which he examines the relationship between 'aspectus' and 'affectus', between the gaze and (divine) love:

> In the same way as light is understood to mean the knowledge of truth, with regard to the *glance* of the mind, in just that way it is understood as the love of the known truth in the *desire* of the mind.[6]

From this relationship between *aspectus* and *affectus*, Grosseteste formulated what Richard Southern describes as a "maxim" of his understanding of the physical (and metaphysical) worlds that in many ways can be traced to the very beginnings of linear perspective:

> The first manifestation of a movement toward perspectivity can be found in the new sense of space in painting, architecture, and the organization of cities. What is common to all these areas is a new coordination of space and a representation that takes into account the position of the spectator and his or her appreciation of the visible unity and beauty of the setting.[7]

For the first time, urban space was experienced as a series of ceremonially linked settings, where directional movement gave cities a new and distinct spatial orientation. This orientation, which accompanied a growing interest in the theological and cosmological perspectives of light and vision during the thirteenth century, even extended to the experience of liturgical space.[8]

Perhaps reflecting this new spatial awareness is what Denery describes as the growing importance of appearance in the Middle Ages:

> People had come to think about themselves primarily in visual terms, in terms of a somewhat amorphous distinction between what appears and what exists . . . Confessional manuals, for example, are full of instructions to ensure that penitents and their sins will be fully revealed to their confessors. Not only does the confessor see the penitent, the penitent is taught to see himself through the confessor's gaze.[9]

The reciprocity between seeing and being seen has significant implications in our understanding of the shift from 'perspectivist optics', and its relationship to the late medieval orientational movement and modes of appearance, to full blown pictorial representation in the Renaissance and beyond.

From *perspectiva naturalis* to *perspectiva artificialis*

To understand the assimilation of perspectivist optics during the early Renaissance, and its transformation into *perspectiva artificialis*, we have to confront the difficult question of what was lost and gained in the transformation while mindful of Vesely's claim that perspective constituted "the first plausible anticipation of modernity." In recent scholarship, this issue has tended to focus on the differing perspectives of Leon Battista Alberti and Nicholas Cusanus:

> Alberti's *De pictura*. . . . helped inaugurate what Heidegger came to call "The Age of the World Picture." "Picture" is understood here as something produced by the subject, something that has its center in and receives its measure from the subject. . . . To confront Alberti with Cusanus is to invite our age, this "Age of the World Picture," to recognise the poverty that shadows its power, to become learned about its ignorance. We refuse that invitation at our peril.[10]

Karsten Harries identifies a number of important distinctions between the 'codification' of perspective in Alberti's *De pictura*, with its evidencing of a 'geometrisation' of space, and Cusanus' 'ontological' treatment of perspective in which the "privileging of mathematics has its foundation not in the nature of things, but. . . . in relation to the nature of human understanding."[11] One way of examining the intersection between optics and perspective at this time is to consider the metaphorical use of optical instruments as a means of 'corrective' visualisation and representation – an issue that was especially relevant to Cusanus. Such an examination requires clarification of the meaning of the term 'instrument'. Unlike the modern definition – as a form of equipment or implement to serve largely practical ends – in medieval and early modern parlance, the utility of *instrumentum* was always conditional upon its symbolic/cosmological bearings.[12]

The study of optics, and optical instruments, was very much in vogue in late fourteenth- and early fifteenth-century Florence, drawing upon the vast body of medieval and Arabic Aristotelian studies of light and vision known at the time.[13] For a period, this interest co-existed with the earliest formulations of pictorial space, providing occasions for creative and intellectual exchange. This interest, however, was not just communicated through textual matter and artistic production but also involved oratory, as we see in the famous sermons of Fra Antonino Pierozzi. Towards the end of his life, Fra Antonino (who became Archbishop of Florence) wrote down his sermons in a four-volume work entitled *Summa Theologia*:

> The most interesting aspect of this monumental work was its emphasis on the act of seeing with the eyes, and his use throughout of optical analogies and metaphors to emphasize his spiritual and moral messages. Although he never referred to optics by its common Latin name, *perspectiva*, there is no doubt Antonino was familiar not only with the work of Peter of Limoges [*De oculo morali* ('Concerning the Moral Eye')] but with the basic principles of the visual pyramid as applied by Bacon to his species theory.[14]

We have to remember that these sermons were delivered in basilicas and public places where Fra Antonino's use of optical analogies and metaphors were conveyed in the very concrete settings of religious/civic spaces. Accordingly, his 'theological optics' would have been received by individual worshippers, each oriented to the altar, and also by assembled congregations in their side-by-side relationships.[15] It is in this context of 'seeing' and 'being seen' (both humanly and divinely) that we can recognise Denery's assertion of the growing importance of appearance in the Middle Ages, in which prayer and confession were central.

Among the audiences of Antonino's sermons would have been artists, humanists and members of the clergy, many of whom contributed to the conception and codification of perspective. It is likely that Nicholas Cusanus would have attended, or known, these sermons and was influenced by Antonino's optical analogies. Cusanus' theology drew upon a deep tradition of Christian-Platonic cosmology, as we see, for example, in his interpretation of the corpus of the pseudo-Dionysius. At the same time, he promulgated radically new ideas regarding his conception of an infinitesimal cosmos that was equated with an infinite/eternal God. These two aspects of his work could be said to converge on his principle of the 'coincidence of opposites', in which the limits of one's capacity to understand divine things (through maximal and minimal relationships) are affirmed through our 'learned ignorance.'[16]

Cusanus communicated his theology of light and geometry in a number of treatises. Of particular interest here is a little work entitled *De Beryllo*, in which a beryl stone is used to convey the 'interfacial' relationships between the act of seeing and mindful vision. It is interesting to note that the choice of this precious stone as a metaphorical instrument in early Renaissance thought was not unique to Cusanus:

> In the discussion of his favourite subject, the refraction of light in semi-precious stones, Ghiberti uses the term "beryl", which does not appear in the literature to which he directly refers – but does figure prominently, as it happens, in the work of Nicholas of Cusa, who dedicated an entire treatise on the subject.[17]

In the case of Lorenzo Ghiberti, his third commentary is a unique study of visual experience by a Renaissance artist which draws upon an extensive literature of medieval optical sources (from Roger Bacon to John Pecham).[18] Along with air, water, glass, crystals and chalcedony, Ghiberti identifies the beryl stone as "diafano . . . per la sua transparentia e rarità."[19] This attribute is reiterated by Cusanus' *De Beryllo* where he states,

> Beryl stones are bright, white, and clear. To them are given both concave and convex forms. And someone who looks out through them apprehends that

which previously was invisible. If an intellectual beryl that had both maximum and a minimum form were fitted to our intellectual eyes, then through the intermediateness of this beryl the indivisible Beginning of all things would be attained.[20]

Later Cusanus observes,

> Let us apply the [intellectual] beryl to our mental eyes, and let us look out through both the maximum (than which there can be nothing greater) and the minimum (than which there can be nothing lesser), and we will see the Beginning, prior to everything great or small.[21]

Cusanus demonstrates the inter-dependency between maximal and minimal by constructing a thought experiment revealed through the beryl stone (Figure 1.1). This takes the form of a reed (of length hj) folded in the middle (i) so that one half (ij) can rotate as a separate length to form an angle in relation to the other length which is left flat (hi): "As long as the line cd constitutes one angle with ca and another angle with cb, neither angle is maximal or minimal. . . . And so, the one angle does not become maximal before the other becomes minimal."[22]

What transpires from this mental exercise is that "it cannot be called an acute angle or a right angle or an obtuse angle, for it is not any such angle but is the most simple cause of [angles]."[23] Instead, "we know that the maximal and the minimal [angle] is the complete totality and perfection of all formable angles and is both the innermost center and the containing circumference of them all."[24] This arrangement of graduating angles is highlighted in Figure 1.1.

Accordingly, through the intellectual eyeglass, we encounter not a static geometric configuration but a geometry of mediation by means of angular rotation that reveals the quiddity of angularity. Significantly, geometry serves here as a similitude of mystical revelation analogous to the efficacy of light. There is, however, no mention of optical magnification in Cusanus' description (in spite of his reference to the concave and convex shapes of the beryl stone). Instead, we are left with the impression of the stone as a pure lens that affords the most elevated vision of a divine order.

In one sense, *De Beryllo* relates to a deep history of the analogical meanings of eyeglasses, but one that transcends empirical demonstration about optical performance. Cusanus takes advantage of the clear and white crystal of the stone to contemplate divine things through geometric relationships, as if the lens is unimpeded by its

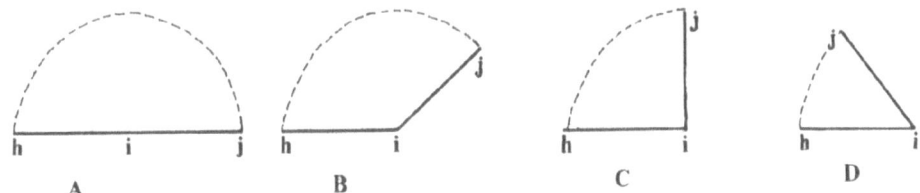

Figure 1.1 Demonstration of "Meditative Angles", *De Beryllo*, Nicholas Cusanus (1400–64), in which a straight line (hj) is the Likeness of True Being (*veritas*) (A), creation of an obtuse angle (at i) as Being (B), right angle as Living (C) and acute angle as Understanding (D)

material nature. The implication of Cusanus' metaphor is that vision of divine things requires the aid of an 'instrument' to reveal what is otherwise hidden.

What *De Beryllo* reveals ontologically, through the agency of animated geometry, is transposed in Cusanus' *De Visione Dei* into a *situational* relationship among participating souls through the senses; the dynamics of the rotating angle, in which maximal is always countered by minimal movement, serves as an analogy to the intersecting (mutually dependent) perspectives of communing worshippers. This relationship is reinforced by Michel de Certeau's description of *De Visione Dei* as a "mathematical liturgy".[25]

Unlike, however, most other tracts in Cusanus' *oeuvre*, in which the cardinal deploys mathematical and geometrical similitudes to demonstrate his ideas, in *De Visione Dei* he "integrates the fundamental exercises of seeing, hearing, and speaking into a praxis designed to lead his monastic audience" into what he calls the "ready access to mystical theology" *(facilitas mysticae theologiae)*.[26] Cusanus deploys a religious icon of Christ to communicate this experience:

> Each monk first stands observing the [its] atemporal gaze seemingly directed to him alone. Then, each moves from his original location to the opposite side, in amazement at the "change of the unchangeable gaze" that introduces temporality and mutability into the experience. The crucial part of the experiment, however, is the transition from the visual realm to the audible, the speaking and hearing that form the believing community – that is, faith is what enables us to begin to move beyond the perspectival as particular toward a more universal viewpoint. The simultaneous omnivoyance, or infinity, of Christ's gaze begins to be revealed only when each brother asks the other as they meet simultaneously.[27]

'Omnivoyance' typically refers to the optical illusion created by a figure painted on a flat surface in which the eyes stare directly out of the picture at 90 per cent to the surface. Bernard McGinn, however, construes the overall effect of this all-seeing presence of the icon as moving "beyond the perspectival as particular toward a more universal viewpoint." The synthesising of the human senses and oral communication provides perhaps the most far-reaching account of a *situational* understanding of the sacred. But can we reconcile Cusanus' 'model' of optical perspective with developments in pictorial representation during this time? This question prompts us to return to a comparison between Cusanian and Albertian concepts of perspective, between Cusanus' contemplative geometries (in *De Beryllo*) and Alberti's demonstrative geometries (in *De pictura*).

Clifton Olds observes in this comparison

> the principle of one point perspective [that forms the basis of Alberti's *De pictura*] runs *absolutely* counter to Cusanus' metaphor, in that the former assumes that the viewer of a painting is always standing in front of it, his eyes on a level with the vanishing point, and at a specific distance from the image. As a demonstration of our own limited – or to use Cusanus' adjective, "contracted" – vision, Renaissance perspective is in itself an ideal metaphor, one that Cusanus could have employed in drawing the contrast between God's all-seeing eye and own barrel vision. It is antithetical to the limitless vision of God, however; and Cusanus would have been the first to see the contradiction.[28]

Olds makes a clear distinction between linear perspective and what he calls "aspective principles": "looking *at* rather than looking *through*."[29] The distinction, however, is challenged by Charles Carmen who argues for a closer relationship between Albertian principles of perspective and Cusanus' 'theology' of vision:

> Cusanus is explicit in his interpretation of observing God's perspective, explaining that upon realizing the miraculous nature of God's omnivoyance, "unless he [a brother] believed, he would not apprehend that this. . . . was possible". . . . The icon's viewer must apprehend, or grasp. . . . , must look into himself using faith and intellect. Only there will he find some explanation for what can be known beyond sense certainty. Alberti, on the other hand, is less explicit about what his infinite geometry transmits, though, as I have suggested, something similar to Cusanus's conception applies to the viewer of these Albertian spaces. . . . Although expressed in different settings their methods indicate a common goal. Cusanus transforms the non-geometrical perspective of viewing the icon into a geometric metaphor of God's actual infinity; Alberti encloses the natural physiology of what we see within a visibly infinite geometry. Both start from simply "natural" appearance and use the metaphoric power of geometry to advance to a higher understanding of that appearance.[30]

In Cusanus' 'contemplative' geometries, understanding one's relationship with God is 'measured' on the basis of one's capacity to seek (but ultimately not attain) knowledge of divine things through the agency of 'geometric reckoning', hence our 'learned ignorance'. Accordingly, Cusanus deploys 'open' geometry to reveal this gap between human finitude and divine infinitude. Alberti, on the other hand (to quote Carmen), "*encloses* the natural physiology of what we see". The principle of enclosure – or framing the visible world – relates back to Harries' argument (quoted earlier in this chapter) that perspective is an anticipation of the Heideggerian concept of the modern "World Picture". Crucially, Harries' reasoning behind comparing the ideas of Alberti to Cusanus is that the former is "one of the founders of our modern world. . . . whose material wealth is shadowed by spiritual poverty".[31] The end result of Albertian perspective – "a visibly infinite geometry" as Carmen describes it – seems to counter Cusanus' geometry as a thought experiment to express the synonymy between infinite space and the infinitude of God.

This comes to a further issue underlying this comparison – namely, Cusanus' abiding emphasis on direct (worldly) experience to *reveal* humanity's relationship to God, which is a principle that is conspicuously absent in Alberti's work.[32] In this outlook, Cusanus rarely considers divine matters in purely abstract terms, since his approach to mathematics and geometry is always mediated through experience in some way. To take an example, in Cusanus' *Idiota* dialogues (*Idiota de sapientia*) Harries observes,

> Having proclaimed, citing Scripture, that wisdom cries out in the streets, the layman calls the orator's attention to the activities that take place in the marketplace: money being counted, oil being measured, and produce being weighed. In each case a unit measure is applied to what is to be measured. And can we not observe something of the sort wherever there is understanding? The activities observed on the marketplace invite the thought that just insofar as he is the being who measures, the human being transcends the beast. *Animal rationale* comes to be understood first of all as *animal mensurans*.[33]

Cusanus' equation of 'measure' with human understanding, as it pertains to the everyday life of the marketplace, reminds us of Certeau's description (referred to earlier) of the 'visual' encounters of communing monks in *De Visione Dei* as a form of "mathematical liturgy". This relationship prompts us to recollect the historical connections between the 'invention' of perspective and architecture. Understanding such a relationship is important if we are to begin to construe how perspective became instrumentalised.

Brunelleschi's experiment

The importance of Brunelleschi's famous perspective 'experiments' in Florence on Alberti's codification of *perspectiva artificialis* in his *De pictura* has been extensively debated by scholars.[34] What is often overlooked, however, in these studies are the symbolic implications of Brunelleschi's use of urban/building fabric to test his method and his choice of sites in the city in which to construct these pictorial projections. As Martin Kemp reminds us,

> The procedures relied upon existing buildings and, inevitably, resulted in the portrayal of these buildings. Painters were not employed to paint townscapes as such, except in very unusual circumstances, and a set of existing buildings is unlikely to have provided an appropriate or adaptable setting for the religious subject-matter which predominated.[35]

Should we accept the principle, therefore, that Brunelleschi's 'discovery' (or invention) of perspective was borne out of the visualisation and reconstruction of the physical urban fabric of Florence rather than formulated in purely abstract terms as an imagined geometrical construct (as we later see in Alberti's *De pictura*)? Early in his career (in 1413), Brunelleschi was called *prespettivo ingegnoso uomo* ('ingenious perspective man') – a term that Samuel Edgerton argues can only refer to optics (or *perspectiva naturalis*) rather than to linear perspective (*perspectiva artificialis*). This is on account of Brunelleschi's association with the charismatic orator, Fra Antonino, referred to earlier.[36] Through his exposition of a 'civic theology', Antonino viewed the city of Florence as a setting for blending "experiential existence with its commitment to salvation, a *mirror image* of the heavenly Jerusalem."[37]

It is appropriate, therefore, that we draw upon this theological background to examine Brunelleschi's famous perspective reconstruction of Florence Baptistery (San Giovanni). It is my intention here only to examine the work in the context of its topographical location. It would be helpful, however, to recapitulate Brunelleschi's perspective methods according to his biographer, Antonio Manetti:

> He first demonstrated his system of perspective in a small panel about a half braccio square. He made a representation of the exterior of San Giovanni of Florence, encompassing as much of that temple as can be seen at a glance from the outside. In order to paint it, it seems that he stationed himself three braccia inside the central portal of Santa Maria del Fiore. . . . In the foreground he painted that part of the piazza encompassed by the eye. . . . And he placed burnished silver where the sky had to be represented, that is to say, where the buildings of the painting were free in the air, so that the real air and atmosphere were reflected in it. . . . he made a hole in the painted panel at that point in the temple of San Giovanni

which is directly opposite the eye of anyone stationed inside the central portal of Santa Maria del Fiore, for the purpose of painting it. . . . He required that whoever wanted to look at it place his eye on the reverse side where the hole is large, and while bringing the hole up to his eye with one hand, to hold a flat mirror with the other hand in such a way that the painting would be reflected in it. The mirror was extended by the other hand a distance that more or less approximated in small braccia the distance in regular braccia from the place he appears to have been when he painted it up to the church of San Giovanni. With the aforementioned elements of the burnished silver, the piazza, the viewpoint, etc., the spectator felt he saw the actual scene when he looked at the painting.[38]

Edgerton makes the case that this experiment was conducted very much under the influence of a "still medieval intellectual ambience of [Brunelleschi's] time", and accordingly "he would not have immediately have conceived of it as a purely abstract geometrical function."[39] In the experiment, Brunelleschi incorporated a combination of pictorial reconstruction (depiction) and 'catoptric certification' (reflection). The relationship between both is significant when we consider the differences between *perspectiva artificialis* and *perspectiva naturalis*, between the penetrative properties of projective geometry and the reflective properties of burnished silver redolent of the gold backgrounds of medieval icons.

The chosen location for Brunelleschi's experiment is likely to have been a motivating force in the project. As the focus of the *quartiere di San Giovanni*, the Baptistery was also regarded as the oldest and most venerated building in Florence, hence Manetti's reference to it as a 'temple' (Figure 1.2). This symbolic and historical importance is

Figure 1.2 View of the Baptistery of San Giovanni taken from the central portal of Florence Cathedral. Akg-images/Album/Prisma

further underpinned by the ceremonial relationship between the Baptistery and the Cathedral opposite it:

> Brunelleschi *cum* artist chose an ideal position from which to view his subject by standing just inside the Cathedral door. Positioned there he views and reproduces the space towards the eastern doors of the Baptistery, which, facing the western entrance to the Cathedral constitute the exit way of the newly baptized who would traverse this space of paradise as the literal path leading to the Eucharistic altar and union with God. His subject, in other words, was the sacred journey this space defined (Figure 1.3).[40]

Hence the simultaneous depiction/reflection of the Baptistery and its surroundings redefined, by optical means, the liturgical connection between the rites of Baptism and the Eucharist. The space between the Baptistery and Cathedral, which had been widened after the demolition of Santa Reparata, and the construction of the larger basilica further east, allowed the whole building of the Baptistery to be visible from the Cathedral portal:

> The east doors faced directly onto the main portal of the Cathedral opposite. When both entrances were open, people in the Baptistery could view not only the newly finished west facade recently adorned with life-size sculptures of the Four Evangelists, but the main altar inside the Cathedral itself, while people in the Cathedral could in turn admire the sacred Baptistery, framed by the Cathedral doorway almost like a holy relic in a tabernacle.[41]

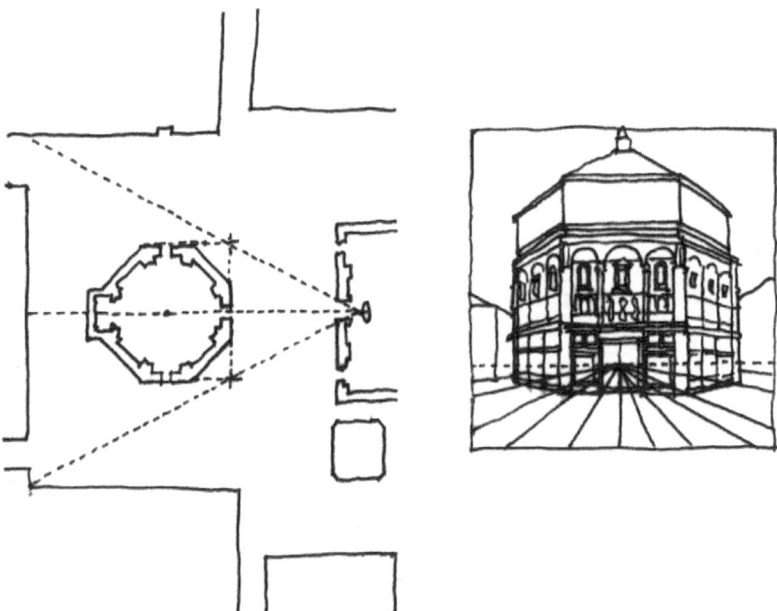

Figure 1.3 Plan of Florence Cathedral and Baptistery (after Bernardo Sansone Sgrilli) showing a hypothetical reconstruction of Brunelleschi's visual angle from central portal of the Duomo to the Baptistery, with perspective view of the Baptistery (drawn by author)

The religious and civic importance of the piazza between the Baptistery and Cathedral, that formed the spatial setting of Brunelleschi's panel, is underlined by its association with 'earthly paradise' as described in Leonardo Bruni's *Panegyric to The City of Florence*, written at the beginning of the fifteenth century.[42] In Brunelleschi's in situ perspectival "certification" of the Baptistery we are reminded again of Certeau's interpretation of Cusanus' *De Visione Dei* as a "mathematical liturgy". Analogies between liturgical space and forms of mathematical and geometric reckoning also find resonance in Marvin Trachtenberg's argument that the Baptistery "was the first link in a single proportional chain that eventually came to structure all three buildings of the cathedral group [new Duomo, Campanile and Baptistery]".[43] Trachtenberg demonstrates this by highlighting the way the dimensions of both Duomo and Campanile derive from the Romanesque Baptistery, as we see in Franceso Talenti's 1357 revision; the external width of the Duomo (72 braccia) comprises the sum of the widths of the three sides of the octagonal Baptistery facing the west facade of the Cathedral and represented in Brunelleschi's perspective projection (24 + 24 + 24 = 72).[44]

Whilst Brunelleschi's panel demonstrated the virtuosity of the new perspective technique, it seems clear that what he revealed in the process was much more; the visual trajectory linking the sites of Baptism (San Giovanni) and the Eucharist (Santa Maria del Fiore) is rendered as 'Paradise' *by virtue* of the site's pictorial reconstruction through optical transformation into an infinite space. In this process, the vanishing point, which Olds speculates "is the perfect analogy to Cusanus' concept of the *finis sine fine*" ('endlessly end'), coincides with the position of the viewing eye (behind the tiny "lentil bean" hole of the painted panel).[45] We are reminded in this analogy of Cusanus' experiential understanding of the 'coincidence of opposites', where "impossibility coincides with necessity."[46] Significantly, Cusanus calls this encounter "the wall of paradise whose gate is guarded by reason."[47]

Brunelleschi's act of 'testing' his pictorial representation of the Baptistery, and its surroundings (the setting of 'Paradise'), both *explicitly* demonstrates what is visible in front and *implicitly* acknowledges what lies behind – the liturgical destiny of the procession of the baptised (the site of the Eucharist).[48] In this 'Janus' relationship, Brunelleschi's mirror and panel constitute as much an *ontological* as a purely *optical* instrument, echoing Fra Antonino's "moralising optics" conveyed in his animated sermons in the religious and civic spaces.

The significance of the translation of Brunelleschi's perspective rendering to the largely abstract codification of pictorial space expressed in Alberti's *De pictura* should not be underestimated. As Kemp makes clear,

> What was needed was a means of adapting Brunelleschi's procedures to the creation of an imagined space [Alberti's *costruzione leggitima*] which could act as the servant to the artist's needs. Without such a means, the potential of the invention would remain dormant.[49]

The term 'dormant' is perhaps unjustified here, as an indication of some expected – but ultimately unfulfilled – potential. Brunelleschi was almost certainly unaware of the full implications of his experiment, given what was to follow in the progressive 'grip' of perspective on both imagined (ideal) and actual (physical) settings.[50] The work, therefore, should not be construed as a conscious demonstration of *perspectiva artificialis* as some have argued (given the lack of supporting geometric procedures), but rather as an affirmation of the still pervasive late medieval tradition of perspectivist optics and catoptrics.

What we see evolving, therefore, in early fifteenth-century Florence is a remarkable and complex series of developments in artistic representation that, initially at least, were founded on a deep tradition of embodied space (Figure 1.4). Brunelleschi's 'experiment' should be understood in this context, in what Vesely calls "the slow perspectivization of the culture as a whole."[51] Whilst the decisive turning point is signalled by Alberti's codified rules of perspective, in which "A God-centred art gives way to a subject centred art", we have to wait until the beginning of the eighteenth century before the full consequences of this invention (or 'rediscovery') are manifested.[52] For the rest is history as we encounter in our digital world via computer screens, tablets and simulated/augmented realities.

Figure 1.4 Detail of right hand side of the fresco 'Raising of the Son of Theophilus and St Peter Enthroned' by Masaccio (1426–27) and Filippino Lippi (1457–1504), Brancacci Chapel, Church of Santa Maria delle Carmine, Florence. Nicholas Temple

Notes

1 Hubert Damisch, *The Origin of Perspective* (Cambridge, MA: MIT Press, 1995), 107.
2 Dalibor Vesely, *Architecture in the Age of Divided Representation: The Question of Creativity in the Shadow of Production* (Cambridge, MA: MIT Press, 2004), 6.
3 This chapter is a further development of an earlier study, Nicholas Temple, *Disclosing Horizons: Architecture, Perspective and Redemptive Space* (London: Routledge, 2007).
4 Dallas G. Denery II, *Seeing and Being Seen in the Late Medieval World: Optics, Theology and Religious Life* (Cambridge: Cambridge University Press, 2005), 4–5.
5 Ibid., 5.
6 C.F.J. Martin (trans.), *Robert Grosseteste: On the Six Days of Creation: A Translation of the Hexaëmeron* (Oxford: Oxford University Press, 1996), Part II, ch. IX, 2.
7 Vesely, *Architecture in the Age of Divided Representation*, 110.
8 See Nicholas Temple, John Shannon Hendrix and Christian Frost (eds.), *Bishop Robert Grosseteste and Lincoln Cathedral: Tracing Relationships between Medieval Concepts of Order and Built Form* (Farnham, Surrey: Ashgate, 2014).
9 Denery II, *Seeing and Being Seen in the Late Medieval World*, 7.
10 Karsten Harries, 'On the Power and Poverty of Perspective: Cusanus and Alberti,' in *Cusanus: The Legacy of Learned Ignorance*, edited by Peter J. Casarella (Washington, DC: The Catholic University of America Press, 2006), 108; Charles H. Carman, *Leon Battista Alberti and Nicholas Cusanus: Towards an Epistemology of Vision for Italian Renaissance Art and Culture* (Farnham, Surrey: Ashgate, 2014).
11 Harries, 'On the Power and Poverty of Perspective,' 109. See also Ronald Godzinski Jr., '(En)Framing Heidegger's Philosophy of Technology,' *Essays in Philosophy*, Vol. 6, Issue 1 (January, 2005), Article 9.
12 Donald G. Bates, 'Machina Ex Deo: William Harvey and the Meaning of Instrument,' *Journal of the History of Ideas*, Vol. 61, Issue 4 (October, 2000), 577–593.
13 Samuel Y. Edgerton, *The Mirror, the Window, and the Telescope: How Renaissance Linear Perspective Changed Our Vision of the Universe* (Ithaca, NY: Cornell University Press, 2009), 30.
14 Ibid., 32.
15 For further discussion on this see Nicholas Temple 'Architecture as the Receptacle of *Mitsein*,' in *Intersections of Space and Ethos*, edited by Kyriaki Tsoukala, Nikolaos-Ion Terzoglou and Charikleia Pantelidou (London: Routledge, 2015), 138–149.
16 Nicholas of Cusa, *On Learned Ignorance: A Translation and an Appraisal of De Docta Ignorantia*, translated by Jasper Hopkins (Minneapolis, MN: The Arthur J. Banning Press, 1985).
17 Vesely, *Architecture in the Age of Divided Representation*, 161.
18 Some have dismissed Ghiberti's third commentary as regressive on account of the lack of discussion of linear perspective. Ibid., 160.
19 "Diaphanous . . . for its transparency and rarity." Alberto Ambrozini, *Immaginazione visiva e conoscenza: Theoria della Visione e Practica Figurativa nei Trattati di Leon Battista Alberti, Lorenzo Ghiberti, Leonardo da Vinci* (Pisa: Edizioni plus – Pisa University Press, 2008), 158.
20 Nicholas of Cusa, 'On [Intellectual] Eyeglasses (De Beryllo),' in *Nicholas of Cusa: Metaphysical Speculations*, translated by Jasper Hopkins (Minneapolis, MN: The Arthur J. Banning Press, 1998), 36(3).
21 Ibid., 37(8).
22 Ibid., 39(10).
23 Ibid., 40(12).
24 Ibid., 41(15).
25 Michel de Certeau, 'The Gaze of Nicholas of Cusa,' in *Diacritics: A Review of Contemporary Criticism*, edited by Jason Hopkins, Vol. 17, Issue 3 (Fall, 1987), 14.
26 Bernard McGinn, 'Seeing and Not Seeing: Nicholas of Cusa's De Visione Dei in the History of Western Mysticism,' in *Cusanus: The Legacy of Learned Ignorance*, edited by Peter J. Casarella (Washington, DC: The Catholic University of America Press, 2006), 38.
27 Ibid., 39.
28 Clifton Olds, 'Aspect and Perspective in Renaissance Thought: Nicholas of Cusa and Jan van Eyck,' in *Nicholas of Cusa on Christ and the Church: Essays in Memory of Chandler McCuskey Brooks for the American Cusanus Society*, edited by Gerald Christianson and Thomas M. Izbicki (Leiden: Brill, 1996), 254.
29 Ibid., 255.

30 Carman, *Leon Battista Alberti and Nicholas Cusanus*, 91.
31 Harries, 'On the Power and Poverty of Perspective,' 108.
32 This is most clearly highlighted in Cusanus' account of the inspiration for his concept of learned ignorance. Harries, 'On the Power and Poverty of Perspective,' 112.
33 Ibid., 119.
34 See in particular Damisch, *The Origin of Perspective*; Martin Kemp, *The Science of Art: Optical Themes in Western Art from Brunelleschi to Seurat* (New Haven: Yale University Press, 1990) and Samuel Y. Edgerton, *The Renaissance Rediscovery of Linear Perspective* (New York: Harper Icon, 1976).
35 Kemp, *The Science of Art*, 15.
36 Edgerton, *The Mirror, the Window, and the Telescope*, 40.
37 Carmen, *Leon Battista Alberti and Nicholas Cusanus*, 125–126.
38 Howard Saalman (ed.), *The Life of Brunelleschi by Antonio di Tuccio Manetti*, translated from the Italian by Catharine Enggass (University Park: Pennsylvania State University Press, 1970), 42–44.
39 Edgerton, *The Mirror, the Window, and the Telescope*, 48.
40 Carman, *Leon Battista Alberti and Nicholas Cusanus*, 126.
41 Ibid., 127. Brunelleschi stood approximately 60 braccia away from the east facade of the Baptistery.
42 Benjamin G. Kohl, Ronald G. Witt and Elizabeth B. Welles (eds.), 'Leonardo Bruni: Panegyric to the City of Florence,' in *The Earthly Republic: Italian Humanists on Government and Society* (University Park: University of Pennsylvania Press, 1978), 141–142. Krautheimer states that the east entrance to San Giovanni was the "focus, as it were, for the entire body of sacred buildings on the border between the old and the new parts of the town." Richard Krautheimer and Trude Krautheimer-Hess, *Lorenzo Ghiberti* (Princeton: Princeton University Press, 1970), Vol. 1, 34. Quoted in Edgerton, *The Mirror, the Window, and the Telescope*, 47.
43 Marvin Trachtenberg, 'Architecture and Music Reunited: A New Reading of Dufay's "Nuper Rosarum Flores" and the Cathedral of Florence,' *Renaissance Quarterly*, Vol. 54, Issue 3 (Autumn, 2001), 749. Trachtenberg makes a compelling argument for a numerical correspondence between Guillaume Dufay's motet, Nuper Rosarum Flores, commissioned for the dedication of the new cathedral in 1436, and the proportions of the Duomo that in turn derive from the Baptistery.
44 Ibid., 750.
45 Old, 'Aspect and Perspective in Renaissance Thought,' 254. Olds asserts, however, that this is his interpretation, not Cusanus'.
46 McGinn, 'Seeing and Not Seeing,' 46.
47 Ibid.
48 As Edgerton states, "In our mortal world, just as in my mirror, you see the Baptistery only enigmatically. Not until you are in heaven face to face with God, will you at last behold its true reality." *The Mirror, the Window, and the Telescope*, 53.
49 Kemp, *The Science of Art*, 15.
50 See Temple, *Disclosing Horizons*.
51 Vesely, *Architecture in the Age of Divided Representation*, 110.
52 Harries, 'On the Power and Poverty of Perspective,' 111.

Bibliography

Ambrozini, Alberto. *Immaginazione visiva e conoscenza: Theoria della Visione e Practica Figurativa nei Trattati di Leon Battista Alberti, Lorenzo Ghiberti, Leonardo da Vinci*. Pisa: Edizioni plus – Pisa University Press, 2008.

Bates, Donald G. "Machina Ex Deo: William Harvey and the Meaning of Instrument." *Journal of the History of Ideas* 61.4 (October, 2000): 577–593.

Carman, Charles H. *Leon Battista Alberti and Nicholas Cusanus: Towards an Epistemology of Vision for Italian Renaissance Art and Culture*. Farnham, Surrey: Ashgate, 2014.

Certeau, Michel de. "The Gaze of Nicholas of Cusa." *Diacritics: A Review of Contemporary Criticism* 17.3 (Fall, 1987): 2–38.

Damisch, Hubert. *The Origin of Perspective*. Cambridge, MA: MIT Press, 1995.

Denery II, Dallas G. *Seeing and Being Seen in the Late Medieval World: Optics, Theology and Religious Life*. Cambridge: Cambridge University Press, 2005.

Edgerton, Samuel Y. *The Mirror, the Window, and the Telescope: How Renaissance Linear Perspective Changed Our Vision of the Universe*. Ithaca, NY: Cornell University Press, 2009.

Edgerton, Samuel Y. *The Renaissance Rediscovery of Linear Perspective*. New York: Harper Icon, 1976.

Godzinski Jr., Ronald. "(En)Framing Heidegger's Philosophy of Technology." *Essays in Philosophy* 6.1 (January, 2005): Article 9.

Harries, Karsten. "On the Power and Poverty of Perspective: Cusanus and Alberti." In *Cusanus: The Legacy of Learned Ignorance*, edited by Peter J. Casarella, 105–126. Washington, DC: The Catholic University of America Press, 2006.

Kemp, Martin. *The Science of Art: Optical Themes in Western Art from Brunelleschi to Seurat*. New Haven: Yale University Press, 1990.

Kohl, Benjamin G., Ronald G. Witt and Elizabeth B. Welles (eds.). "Leonardo Bruni: Panegyric to the City of Florence." In *Earthly Republic: Italian Humanists on Government and Society*, translated by Benjamin G. Kohl, 135–177. University Park: University of Pennsylvania Press, 1978.

Krautheimer, Richard and Trude Krautheimer-Hess. *Lorenzo Ghiberti*. Princeton: Princeton University Press, 1970, Vol. 1.

Martin, C.F.J. (trans.). *Robert Grosseteste: On the Six Days of Creation: A Translation of the Hexaëmeron*. Oxford: Oxford University Press, 1996.

McGinn, Bernard. "Seeing and Not Seeing: Nicholas of Cusa's De Visione Dei in the History of Western Mysticism." In *Cusanus: The Legacy of Learned Ignorance*, edited by Peter J. Casarella, 26–53. Washington, DC: The Catholic University of America Press, 2006.

Nicholas of Cusa. "On [Intellectual] Eyeglasses (De Beryllo)." In *Nicholas of Cusa: Metaphysical Speculations*, translated by Jasper Hopkins, 35–72. Minneapolis, MN: The Arthur J. Banning Press, 1998.

Nicholas of Cusa. *On Learned Ignorance: A Translation and an Appraisal of De Docta Ignorantia*, translated by Jasper Hopkins. Minneapolis, MN: The Arthur J. Banning Press, 1985.

Olds, Clifton. "Aspect and Perspective in Renaissance Thought: Nicholas of Cusa and Jan van Eyck." In *Nicholas of Cusa on Christ and the Church: Essays in Memory of Chandler McCuskey Brooks for the American Cusanus Society*, edited by Gerald Christianson and Thomas M. Izbicki, 251–264. Leiden: Brill, 1996.

Saalman, Howard (ed.). *The Life of Brunelleschi by Antonio di Tuccio Manetti, Translated from the Italian by Catharine Enggass*. University Park: Pennsylvania State University Press, 1970.

Temple, Nicholas. "Architecture as the Receptacle of *Mitsein*." In *Intersections of Space and Ethos*, edited by Kyriaki Tsoukala, Nikolaos-Ion Terzoglou and Charikleia Pantelidou, 138–149. London: Routledge, 2015.

Temple, Nicholas. *Disclosing Horizons: Architecture, Perspective and Redemptive Space*. London: Routledge, 2007.

Temple, Nicholas, John Shannon Hendrix and Christian Frost (eds.). *Bishop Robert Grosseteste and Lincoln Cathedral: Tracing Relationships between Medieval Concepts of Order and Built Form*. Farnham, Surrey: Ashgate, 2014.

Trachtenberg, Marvin. "Architecture and Music Reunited: A New Reading of Dufay's 'Nuper Rosarum Flores' and the Cathedral of Florence." *Renaissance Quarterly* 54.3 (Autumn, 2001): 740–775.

Vesely, Dalibor. *Architecture in the Age of Divided Representation: The Question of Creativity in the Shadow of Production*. Cambridge, MA: MIT Press, 2004.

2 Desargues' *oeuvres*

On perspective, optics and conics

Nader El-Bizri

Preamble

This chapter examines the visioning techniques in the perspectival and projective geometric methods of Girard Desargues (*Sieur Girard Desargues Lyonnois*; d. 1661). This line of inquiry is undertaken in view of investigating his aims in aiding the construction of architectural space through the mathematical underpinning of its visual representation with greater precision and objectivity. Desargues built on techniques in optical knowledge that he and his contemporaries inherited from the *perspectiva* traditions of the Italian Renaissance in addition to furthering the development of geometry in its projective modes as an inquiry concerning the invariant properties of figures and solids that are subjected to geometrical transformation, with special attention given to conics and gnomonic projections (the latter dealing with plotting spherical surfaces on a rectilinear plane in geodesic mapping and designing sundials). Desargues undertook these investigations in the context of his scientific research as a geometer and optician besides his technical knowledge as a practicing architect who concentrated on the arts of masonry and stonecutting (*la coupe des pierres*). The projective methods he deployed in establishing his perspective constructions were laid down on geometrical optics and posited a convergence point at infinity that opened up Euclidean space beyond what his predecessors in the Renaissance did in their take on the perspectival *costruzione legittima* (Brunelleschi, Alberti, etc.). As an architect, Desargues made contributions to rigorous and positive scientific knowledge. His architectural inquiries expanded the realm of understanding natural phenomena through research in the exact sciences and by way of adopting their mathematized epistemic and methodological directives in serving surveyors and engineers to perfect their mensuration.

Desargues' scientific *milieu*

Desargues' theoretical works in pure mathematics focused on the study of conics and investigating the mathematical and optical principles that grounded perspective.[1] These inquiries, along with his systematized development of projective geometry aimed at a small circle of authoritative scientific readers,[2] while his vernacular epistles were addressed to craftsmen, builders, artisans as promoted by the learned printmaker and engraver Abraham Bosse (d. 1676).

One of Desargues' important studies in conics and projective geometry was the *Brouillon Projet d'une atteinte aux événements des rencontres du cône avec un plan*.[3] This consisted of a draft projective inquiry in conics that focused on the outcome of intersecting a cone with a plane. Desargues discussed in this epistle the properties of conic sections that remain invariant under projection and offered a unified theory of conics in the Apollonian tradition. Even though Descartes had reservations about this *Brouillon*, judging it as being written in an obscure language that borrowed signs from the code-markers of masons and carpenters, he nonetheless acknowledged the impact it had on Pascal's *Essai pour les coniques* (Paris, 1640).[4]

The *oeuvres* of Desargues as a geometer and architect embodied the culmination of long-standing inquiries in mathematics and the exact sciences that preoccupied erudite architects since the early periods of the Renaissance, which were principally underpinned by classical Ancient Greek, mediaeval Latin, and Arabic sources. The Renaissance *perspectiva* legacies that inspired Desargues' research were not only rooted in theoretical and practical investigations in architecture, or the visual and plastic arts, but they were also closely entangled with the mathematical determinants of optics. These Renaissance traditions were grounded on the transmitted and adapted knowledge that was derived from European scholastic mediaeval sources and their assimilation of Ancient Greek and Arabic legacies in science and philosophy. The unfolding of the *perspectiva* traditions in Renaissance architecture, the visual and plastic arts, rested in part on prolongations of the scientific research of mediaeval European opticians of the calibre of the Bishop of Lincoln, Robert Grosseteste, and of Franciscan opticians such as Roger Bacon (at the Oxford college), John Peckham, and Erazmus Witelo. These mediaeval scholars were inspired by various Greco-Arabic studies on optics and conics, along with their application in the perfection of scientific instruments, including the geometric modelling of mirrors and lenses, and the refinement of astrolabes and compasses. Arabic traditions in these domains were conducted in the context of experimentation (*i'tibar*) in view of testing the hypotheses of Greek sources on the nature of vision, light, and colour, and they offered critical re-definitions of central notions in geometry and optics of authorities such as Euclid, Archimedes, Apollonius of Perga, Pappus of Alexandria, and Ptolemy.

The most significant contribution to optics from the times of Ptolemy to those of Kepler was embodied in the *Kitab al-Manazir* (*Book of Optics*; known in Latin as *De aspectibus* or *Perspectiva*)[5] of the eleventh-century polymath Ibn al-Haytham (Alhazen; d. ca. 1041 CE). His optical inquiries were based on his experimental method and the mathematical treatment of the notions of classical physics cum natural philosophy, including his geometrization of place, which he conceived as 'imagined qua postulated void' (*khala' mutakhayyal*), and its definition as a spatial extension.[6] His legacy belonged to the Archimedean-Apollonian tradition in Arabic science, which ultimately influenced the European scholastic and Renaissance *perspectiva* traditions[7] and its conceptual prolongations reached the research of Desargues, albeit being framed within a novel early modern epistemic setting.

Dioptrics and conics

Desargues' conics belonged to a long-standing Greek-Arabic, Archimedean-Apollonian legacy in modelling lenses in dioptrics through the use of conic sections (ellipse, parabola, hyperbola) that do not result in aberrations as in spherics. This research was not simply a specialized branch of pure geometry as inherited from Apollonius of Perga (d. ca. 190 BCE)[8] or Pappus of Alexandria (d. 350 CE) but mainly focused on the anaclastic properties of conics. Based on Desargues' studies on conic sections, the parabola was an optimal geometric model in dioptrics and catoptrics that required two geometric elements to plot its curvature: the *focus*, as a fixed point within the concavity of the parabola, and the *directrix*, as a given line that does not pass through the *focus* and is on the other side of the parabolic curve. The parabola becomes the locus of all the points that determine its continuous tracing to infinity on its two sides, such as the distance from these points to the focus is equal to the distance that separates them from the directrix. The vertex of the parabola is a point that is at a mid-distance between the focus and the directrix, wherein the straight line connecting the focus to the vertex falls perpendicularly on the directrix, and the perpendicular that connects the focus and vertex is its axis of symmetry, with the *latus rectum* as the chord that is parallel to the directrix and passes via the focal point. So if the distance between the focal point and the vertex is equal to *a*, then the distance from the vertex to the directrix is also equal to *a*, and the width of the *latus rectum* is 4*a*; whence $y = ax^2 + bx + c$.

Projective geometry

Desargues' projective theorem in geometry states, *when two triangles are in perspective, the meets of corresponding sides are collinear*. To illustrate (Fig. 2.1), let us have two perspective *homologic* triangles that are not co-planar, ABC and A'B'C', and let their corresponding vertices lie on three distinct straight lines, AA', BB', CC', which all meet at the *perspector* point O, if and only if the respective points of intersection of the corresponding sides of the triangles are co-aligned, such as (BC) ∩ (B'C') = P, (AC) ∩ (A'C') = Q, (AB) ∩ (A'B') = R, and wherein PQR is the co-alignment straight line.

Besides being informed by artificial linear central perspective and the classical science of optics that underpinned it, projective geometry rested also on the reception of the legacy of Pappus of Alexandria (fourth century CE) in conics. Desargues' theorem can be seen as being an exercise in perspective that also had applications in sciagraphy

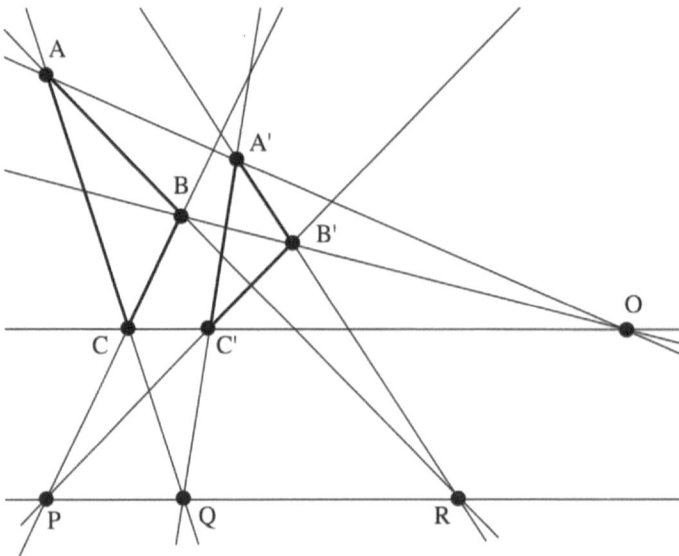

Figure 2.1 The two triangles of Desargues' projective theorem. Image by author

in the projection of shadows in drawing[9] and carried resonances with the antique lemmas of Pappus,[10] as well as Pascal's theorem in the *Essai pour les coniques*.[11]

Desargues' projective geometry studied the invariance of the properties of geometric entities when subjected to transformations (similitude, translation, homothety, affinity) while taking perspective into account as a mode of representing the projected space. This involves co-lineation as a form of homography – namely, by having an isomorphic bijection as a point-by-point correspondence. This allows projections to take place not only on rectilinear surfaces but also on irregular and curved ones as well. Angles in projective geometry are not the same as in Euclidean geometry, since in projective space, they are not invariant as they appear in perspective. For instance, a square appears in variegated trapezoidal forms in perspective as a projective space, while in Euclidean space, they cannot but appear as an invariant square. The angles under projective transformation do not remain the same, and lines in depth tend towards a convergence at the horizon line in what is taken to be a vanishing point at infinity and not simply the vertex of an inverted cone of vision. The projective surface is the picture plane that receives the projections in two-dimensional figures that represent a depiction of projective spatial depth graphically/pictorially. The projective surface can be seen as an extended Euclidean plane. Projective geometry allows the imagination to be active in probing the properties of forms in geometry rather than being strictly an analytic science. We shift herein from a purely mathematical domain to an extension of it that is closer to the realm of embodied experience in visual perception, which lets us imagine the inhabiting of projective space rather than simply standing outside it.

The '*géométral*'

To situate Desargues' *oeuvres* in the context of visioning techniques, we evoke herein his account of what he names '*le géométral*'.[12] This describes a method of drawing that involves the tracing of an orthogonal projection of a given object on a horizontal plane that normally corresponds with the ground, or at times that is set on a vertical plane. The architectural drawing of the *géométral* consists of establishing orthogonal views in a rudimentary form of descriptive geometry, which facilitate the building activity and ease the reading of measurement in the pictorially depicted object. By way of metonymy, the *géométral* designates the plane unto which the projection is made.[13] It is a method of point-by-point bijective correspondence between the set of points assumed to be on the actual object of vision, which is geometrically depicted, and the set of points that are supposed to appear on a squared grid that receives the scaled pictorial depiction of that object: *for every point in a set F of the received projections, there exists one and only one corresponding point in a set E that is being projected: $\forall y \in F, \exists x \in E, f(x) = y$.* This process results in a geometrical description that is axonometric, and hence that offers an objective/positive presentation of objects. Such procedure ensures precision in engineering, surveying, and manufacturing. As for the act of practicing perspective, it consists of drawing the object of vision as close as possible to the way it appears in natural vision by way of a pictorial representation that is guided geometrically. This takes into account the manner by which the seen object appears from a given height, angle, and distance.

Compasses and scaled measurement rulers assist in the use of Euclidean geometry to establish the *géométral* on a marked grid of squares scaled in a single *petit pied* (small step), while the perspective method turns the grid of squares into an interlocked configuration of trapezoidal figures that recede into the projective postulated space in *anamorphosis*, even if the difference between the *géométral* technique and that of perspective is based on the distinction between the trellis cum matrix each uses. However, the *géométral* was conventionally associated with builders, masons, and practitioners of mechanical science (*artes mechanicae*),[14] while perspective was connected with the fine arts of architecture as inherited from the praxis of the Renaissance. Gathering these two methods into congruence gives a nobler status to the *métier* of builders and grants it an architectural status that involves associations with science, the arts and letters. Handworks are henceforth construed as being nearly on par with the fine arts. This also means that gravure and the fine art of painting become co-entangled via the new form of engraving as *portraiture*. Engravers become akin to artists and master-builders are more closely affiliated with architects as proto-engineers, as rendered clear in Bosse's *Manière universelle de M. Desargues pour pratiquer la perspective par petit pied, comme le géométral.*

Perspective

Desargues' projective method in perspective posits a single verging/centring point at infinity through which a visual representation can be established with the same level of accuracy as what in more recent times was achieved in perspectival projections via two vanishing points. The method of projection he uses relies on locating various points in geometric space by plotting a rudimentary coordinate system that determines their location on the floor plan and by being scaled through heights viewed in perspective from the place in which these points are situated. This necessitates an accurate demarcation of the floor plan with its scales and the setting of the various elevations according to their apparent heights within the geometric space of the perspectival representation. Such steps free the pictorial format from relying on frontal elevations that are orthogonal in retaining their limits as surfaces in right-angled intersections, with only the parallel lines in spatial depth converging in the centring point. Rather, surfaces can now be depicted in perspective without the need to always situate some of them in frontal positions. The effect is as accurate as the one accomplished via later perspectival projections that posit two vanishing points, even though the geometric model is still that of a central linear perspective, albeit with its surface depictions standing in the pictorial spectacle under any viewing angle and not solely frontally. A basic aspect of this classic method is shown in Fig. 2.2, which I constructed to show how a given surface '*a*' with a height '*h*' as seen in a plan view X is projected as a surface '*A*' that is viewed in perspective from a height '*d*'.

The use of perspective becomes pivotal to Desargues' endeavour to serve architecture via a scientific mathematical procedure, which also informs the techniques of construction and is not simply a method of organizing architectural space in design, or merely a process of pictorial cum artistic representation. A breakthrough is achieved through the application of perspective in projective geometry, which exceeds the architectural technique of visioning by guiding stonecutters and engravers through mathematical mensuration to reach greater precision in design and execution.

Desargues' perspectivism supports architectural construction in applied technical and tectonic terms, as well as aiding the process of pictorially representing architectural space in a conic projective perspective. These pathways ushered novel directions

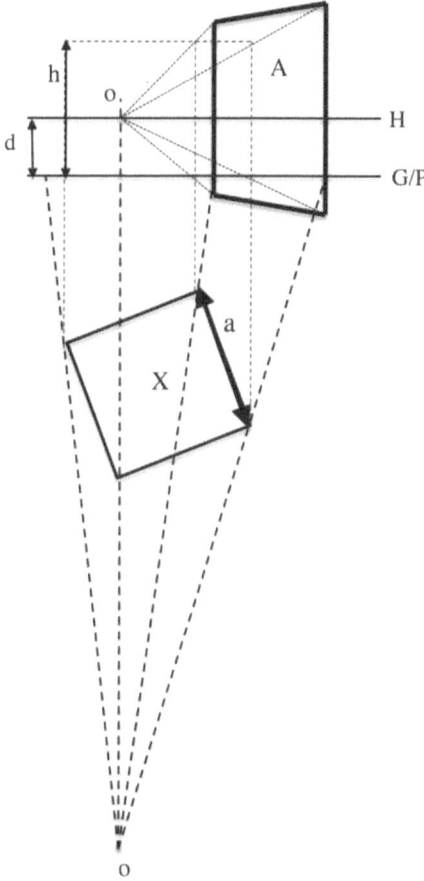

Figure 2.2 Illustration of perspective projection. Image by author

in geometry and its association with optics. Technical practices were expanded through the unfolding of the exact sciences and in their turn aided the development of theoretical science through their applications. Artificial perspective was founded on classical optics, and its applications facilitated the development of projective geometry. The techniques of visualization were not restricted to art and architecture, but they expanded to the domains of projective and descriptive geometry, which also further enhanced the architectural visioning technology. This state of affairs places an epistemic and scientific emphasis on technique and the elevation of building and craftsmanship to a status that is no less noble than the fine pictorial and plastic arts, or architectural design. Bosse already anticipated the conception of the crafts as a hub of technological proto-engineering prior to the European industrial revolution (ca. 1750s–1850s) as exemplified in his reception of the *opusculum mathematicum: Exemple de l'une des manières universelles du S. G. D. L. touchant la pratique de la perspective sans emploier aucun tiers point, de distance ny d'autre nature, qui soit hors du champ de l'ouvrage.*[15] This aimed at establishing perspectival projections using geometrical operations that do not involve measurement with rulers or plumb lines. As a geometrician, Desargues begins by placing his definition of technical terms. The words 'perspective', 'appearance', 'representation', and 'portrait' designated the same thing for him (folio 1); similarly, the verbs 'represent', 'portray', 'find the appearance', and 'place in perspective' all signified the same thing. This is also the case with 'levels' and the situation of 'being parallel to the horizon' (hence not converging in the vanishing/centring point) and with being 'perpendicular' as a 'line squared with the horizon' (or to whatever level line that is parallel horizontally to it). The picture plane, or 'tableau', is also defined as a section plan and rectilinear surface upon which a pictorial perspective is established. The face of the rectilinear plane that is exposed to the eye of the observer is the front (*le devant*) of the tableau and the one that is hidden is its back (*le derrière*). As for projections, these are 'straight lines' (folio 2) corresponding with the rectilinear propagation of light in a homogeneous transparent medium.

In this practice of perspective, a single eye is posited as what demarcates the two actual eyes of the observer, without this pushing Desargues into taking positions of whether he adopts an emission theory of visual rays or an intromission one. Since antiquity, these would have been the two main competing hypotheses about the nature of vision and the comportment of physical light, even though Ibn al-Haytham resolved this dispute by way of his geometrical modelling of the cone of vision in explicating the introduction of light into the eyes. However, the classical intromission theories of vision that predated Ibn al-Haytham, which would have been inspired by an Aristotelian natural philosophy qua physics would argue that the form of a visible object is introduced into the eye as abstracted from its matter when the transparent body (such as air) between the eyes of the observer and the objects of vision is actualized as a transmitting medium by physical illumination. Such physical thesis lacked the geometric model that supported it. As for those who were inspired by Platonist, Euclidean, and Ptolemaic mathematical orientations, they upheld an emission theory of light according to which visible objects are seen by way of light rays that consist of non-consuming gentle irradiating fires, which are subtly emitted from the eyes in the shape of a cone or a pyramid of vision. What concerned Desargues was the geometric model that structures the projections of straight lines as they connect to the posited location of the presupposed position of the eye of the observer in a single-point central linear perspective (folio 2). The measurements have to be placed unto the ground plan of the spectacle or subject matter of the artificial perspective (*assiette du sujet*; seat of the subject matter) and the elevations, along with the distances from the demarcated position of the eye of the observer to the picture plane (folio 3). This projective method allowed Desargues to locate any point in geometric space based on the measurements that pertain to the position of that given point on the ground plan and at a given distance from the eye of the observer while being at a given height. This turns perspective into a mensuration instrument that is based on geometric demonstration (folio 10) within an art that is founded on the entanglement of physics with geometry (*en partie de géométrie et en partie de physique*), which as stated by him was not until then explicated in any public book in France: '*ne se trouve en France encore expliquée en aucun livre public*' (folio 10). However, the classical optical research of Ibn al-Haytham already established this isomorphism between physics and geometry in the context of experimental controlled testing, and such an epistemic turn was well documented in the Latin reception of his *Book of Optics* in the European milieu of mediaeval Franciscan perspectivism and the Italian Renaissance.

Desargues advocated the geometrization of the eye by taking its position to be demarcated by a geometric point through which straight lines pass in projection as the point of centring and convergence that tends to infinity. Any straight line that passes through that centring convergence point is named 'the line of the eye': *'ligne de l'oeil'* (folio 11). When the straight lines of the subject matter being depicted are parallel to one another but not parallel to the plane of the tableau, then these converge into the centring point that demarcates the geometric position of the eye of the observer on the plane of the tableau and would be lines of the eye (folio 12).

The figure that follows (Fig. 2.3) is based on my own reconstruction of a perspectival projection that is drafted on the frontispiece folio of Bosse's edition of Desargues' *Exemple de l'une des manières universelles . . . touchant la pratique de la perspective.* On the upper right side of the projection, we have the plan view that locates the *assiette* that is being depicted with the distance that separates it from the position of the observer and with all the elevation scales noted in the drawing for ease of reference (*échelles faites au tableau*). Another scaling device is also constructed within the perspectival projection field along five horizontal lines with a measuring ruler set on line

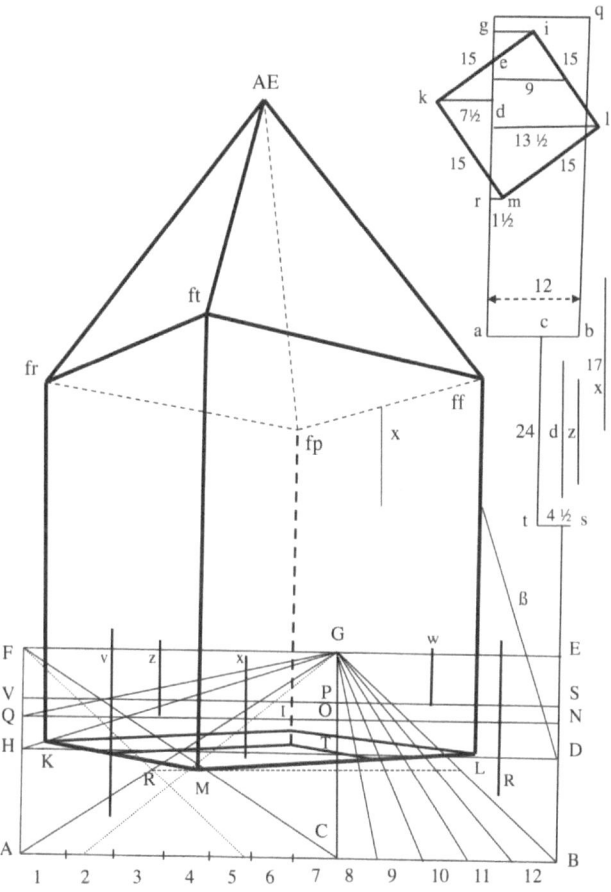

Figure 2.3 Illustration of Desargues' perspective projection. Image by author

ABC that is divided into twelve equal units. The centering/convergence point is located at the horizon line at a point G that is perpendicularly elevated from line ABC at point C, which is situated at seven equal units away from A and at five equal units away from B. The *assiette* is dug within the ground level (at points K, I, L, M) and through it the pictorial projection of the built structure with a pointed roof is elevated.

A marginal note on colour

Desargues' optical research was mainly set in view of grounding perspectivism on mathematics and achieving higher precision in architectural construction. This technical orientation entailed that he was not so much concerned with embodied visual perception or with the study of colour. This aspect would have compromised Bosse's attempt to argue in favour of Desargues' techniques for painting with his own peers in the academy of fine arts. Although Bosse was a founding member of the *Académie Royale de Peinture et de Sculpture* in Paris, his adherence to Desargues' methods, as evident in his pedagogic *Traité*,[16] resulted in clashes with Charles Le Brun (d. 1690). After all, Le Brun passionately promoted emotion and talent rather than acquired technique in the fine arts, as later compiled posthumously in his *Méthode pour apprendre à dessiner les passions* under the influence of the teachings of Nicolas Poussin (d. 1665) within the French classical baroque style. The issue of colour would have been key in approaching the assessment of the fine qualities of painting, and it was also narrowly connected with optics in response to Aristotle's meteorology, as set in the research of figures such as Ibn al-Haytham, Grosseteste,[17] and Roger Bacon.[18] For instance, Grosseteste held that the rainbow concerned the optician qua perspectivist and the physicist qua natural philosopher (*et perspectivi et physici est speculatio de iride. Sed ipsum quid physici est scire, propter quid vero perspectivi*).[19] However, he noted that while the optician/perspectivist seeks explications, the physicist focuses on natural facts. As an optician, he posited colouration as a phenomenon of light that interacts with the nature of transparent media. Colours were not ontologically distinct from light, like it was, for instance, the case with Ibn al-Haytham's speculations in this regard that were later reformed in the research of fourteenth-century opticians such as Kamal al-Din al-Farisi and Theodoric (Dietrich) Teutonicus of Freiburg. Both modelled the raindrop experimentally in a *camera obscura*, whereby they represented it in an enlarged scale in the form of a clear spherical glass vessel filled with water and subjected to controlled rays of light that refracted through it and reflected within it. This experimental inquiry allowed them to examine the geometric structure of the double refractions and reflections of light rays as they passed through the spherical model. Both arrived independently at their similar findings, and each worked separately on manuscripts of Ibn al-Haytham's *Optics*[20] (Arabic in al-Farisi's *Tanqih al-manazir* [*Recension of Optics*][21] and Latin in Theodoric's *De iride et radialibus impressionibus*).[22] The research on the refractive properties of the sphere was based on Ibn al-Haytham's own studies as founded on the dioptrics of the tenth-century mathematician Abu al-'Ala' ibn Sahl who discovered a principle akin to what is commonly known since the seventeenth century as 'Snell's law of refraction' (namely, a principle attributed to Willebrord Snellius that determines the refractive index of a transparent medium in connection with a given geometric shape, which can act as a basis for designing lenses).[23] To better illustrate this principle, if we suppose that n_1 is the refractive index of air and n_2 is the refractive index of water, then $n_2 > n_1$, and the incidence angle Θ_1 would be larger than the refraction angle Θ_2 (both measured against the normal to the surface at the point of incidence and refraction); then $n_1 \sin \Theta_1 = n_2 \sin \Theta_2$.

Proto-engineering art

As noted earlier, Bosse recognized the significance of Desargues' techniques in lifting the *métier* of printmaking, etching, engraving, stonecutting (*coupe de pierres*; *l'art du tailleur de pierre*), carpentry, and instrument making to a status that is possibly as noble as that of the fine arts and architecture. This offered an opportunity to associate handiworks with the letters and the exact sciences. However, such proclaimed elevation of the work of artisans was contested by fine artists at the academy of art in seventeenth-century France. Strife broke in the face of giving scientific legitimacy or artistic license to what in French workmanship is known as '*compagnonnage*' – namely, an apprenticeship lineage that prolongs the method of mentoring and transmission of techniques by habituation within networks of trades and crafts (*les métiers*). This phenomenon designated cross-generational associations that are formed around the construction of large public edifices that took extended periods of labour to be constructed and involved a situational transmission of knowledge, travel, and solidarity, as well as an esoteric preservation of the arcana of the trade.[24] The emancipatory enabling of artisans and craftsmen through the agency of mathematics, optics, and the letters was itself a crossing over of the elite/class divide that placed the painter and the architect at higher ranks.

One craft that becomes an applied proto-engineering science is stonecutting and assembly, which is informed by stereotomy ('*stéréotomie*') that produces sections through solids. This technical knowledge is required in the construction of complex vaults, domes, arches, pillars, cornices, and articulate plastic forms that impose various intersections of solids differing in geometric shapes and demands precision in cutting and assembling stones to fit properly in interlocking outlines. This also involves the art of tracing (*art du trait*) that establishes the scaffolding for designing vaults and arches and acts as an execution drawing drafted to scale.[25] For instance, Eugène Viollet-le-Duc (d. 1879) noted in his *Dictionnaire raisonné de l'architecture française du XIe au XVIe siècle* that '*le trait est une opération de géométrie descriptive, une décomposition des plans multiples qui composent les solides à mettre en œuvre dans la construction*' ('tracing is an operation of descriptive geometry, [namely] a decomposition of the multiple planes that compose solids in order to put them to use in construction').[26] In defining the key geometric entities in stonecutting (*coupe des pierres*), Desargues describes the '*essieu*' as the axis of the vault, the '*sous-essieu*' as the projection of the *essieu* on a facing plane, the '*traverssieu*' as the straight line in the facing plane that is perpendicular to the '*sous-essieu*', and the '*contre-essieu*' as the same plane that cuts through the *essieu* and *sous-essieu*, and is perpendicular to the *essieu*.[27]

The projective techniques of Desargues reach their heights with the architect Amédée-François Frézier (d. 1773)[28] and the geometer Gaspard Monge (d. 1818)[29] and facilitate the systematization of descriptive geometry in the method of *traçage*, whereby a planar rectilinear surface is wrapped, folded, or rolled to generate conical, cylindrical, semi-spherical, prism, and polyhedral shapes. The art of making becomes entangled with the geometrical applied sciences. Drawing allows for a graphical geometrical resolution of the problems of the intersection of complex solids with one another, the tracing of the outlines of their intersections, and setting their orthogonal representation. It also deploys the tracing of curves that are circular, elliptical, parabolic, hyperbolic – which are at the basis of conical – cylindrical, and spherical solids while presupposing various projections on shifting planes in geometric space (top, bottom, front, back, sides). The projective technique in this context consists of a geometric procedure of tracing from any given point in space a straight line that falls as a perpendicular on a rectilinear plane that faces that point at some distance in the Euclidean geometric space.

Stonecutting and assembly, as grounded on stereotomy, embodied techniques that became more precise by way of using geometry in its projective and descriptive methods, with the former branch in geometry implicating linear perspective and its classical optical presuppositions. Such techniques divide solids into smaller ones that are assembled together and determine the outlines of their intersections as they are fitted again together in spatial configurations. These can have architectonic applications, such as curving staircases (helix, helicoids); shaping chimneys; intersecting cylindrical, conical, or spherical sections with masses that have polyhedral forms; or designing complex vaults and partitioning stones that make up domes, arches, etc.

Arts and sciences

Bosse's reception of Desargues' *oeuvres* in a context that promotes the proto-engineering merits of craftsmanship is a mode of classifying and canonizing knowledge in its theoretical and practical domains as part of a process that pre-occupied scholars since antiquity. For instance, the tension between the skills of making, the sciences, and arts and letters was addressed in the architectural milieu through Vitruvius' distinction between *fabrica* and *ratiocinatione*. In the *De architectura*, he argued that

> *fabrica* is the continuous and regular exercise of employment where manual work is done with any necessary material according to the design of a drawing; while *raciocinatione*, on the other hand, is the ability to demonstrate and explain the productions of dexterity on the principles of proportion.
>
> (Vitruvius 193, *Liber* I, *Caput* 1, §1)[30]

However, whilst Vitruvius distinguished between these two spheres, he nonetheless noted that the act of making is not simply a repeated action of the hand but is rather meditative too ('*fabrica est continuata ac trita usus meditatio*').

In *Book VII* of the *Republic*, Plato's curriculum for the would-be philosopher-king included arithmetic, geometry, astronomy, and music qua harmonics (*Politeia*; *De re publica*, 522c-534d).[31] Anicius Manlius Severinus Boëthius (480–524 AD), in paraphrasing Nicomachus of Gerasa's *De institutione arithmetica*, rendered these Greek *tessares methodoi* ('four ways') in Latin as '*quadrivium*'. The '*trivium*' was coined later in the Carolingian ninth century to be combined with the *quadrivium* to complete the seven liberal arts.[32] There were cases in which these were stretched to nine instead of seven. Marcus Terentius Varro (Reatinus; 116–27 BC) listed these in his *Nine Books of Disciplines* (*Disciplinarum libri IX*) as grammar, logic, rhetoric, geometry, arithmetic, astronomy, music, medicine, and architecture. Such classification was echoed in Augustine's *De ordine* (2.12.35 and 2.20.54) and by Martianus Minneus Felix Capella (fl. 410–420) in the *De nuptiis Philologiae et Mercurii* ('Marriage of Philology and Mercury'), which is also entitled *De septem disciplinis* ('The Seven Disciplines').[33]

Therein, Mercurious is associated with eloquence, profitable pursuits, and divination and Philologia with a penchant for the letters. Their wedding gifts were the seven liberal arts offered as maidens to Philologia, while architecture and medicine as earthly arts were present in silence without being included in the *enkuklios paideia* ('education circle') of such *liberalia studia*.

The grouping of architecture with the liberal arts was a rare privilege to accord to it as a field of practice and theory. Classificatory systems carried epistemic and institutional implications in the transition from the high Renaissance to the seventeenth century. The roles that the visual/plastic arts and architecture played in assimilating the applied sciences, the mathematical disciplines, and letters through an effort to study nature in a Renaissance epoch that witnessed the deconstruction of Aristotelian physics all were giving way to novel specialized mathematized models in early modern exact science. This entailed that architecture needed higher precision in its technical applications, which required mathematization. Even though Bosse's promotion of Desargues' techniques compromised his position in the academy of fine arts, the merits of his views found later expressions in engineering. Two of the disciples of Gaspard Monge at the *École Polytechnique* rediscovered Desargues' projective geometry and benefited from its combination with descriptive geometry.[34] Desargues' *oeuvres* were then integrated in the curricula of the *École royale des ponts et chaussées* (established in 1747 in Paris) and the military engineering *École Polytechnique* (set up in 1794, with Monge as one of its founders). Desargues' works also figured in academic journals dedicated in the nineteenth century to military engineers in the French Army.[35] The reception of Desargues' visioning techniques by Bosse signaled that stonecutters and assemblers, carpenters, and engravers could elevate their *métiers* to the status of proto-engineering. This epistemic turn already pointed to what later underpinned industrialization and the unfolding of the essence of technology, which also brought forth the emergence of advanced mathematical techniques in imaging reality and modeling nature by *en-framing* it, or possibly reshaping it.[36]

Notes

1 Girard Desargues, *Oeuvres de Desargues*, ed. Noël-Germinal Poudra (Paris: Leiber, 1864), pp. 11–52. For commentaries on Desargues' contributions in mathematics and the exact sciences refer to Judith V. Field and J. J. Gray, *The Geometrical Work of Girard Desargues* (Dordrecht-Berlin: Springer, 1987); and Girard Desargues, *L'oeuvre mathématique de G. Desargues: Textes publiés et commentés avec une introduction biographique et historique*, ed. René Taton (Paris: Presses Universitaires de France, 1951).

2 Desargues was a member of the mathematical circle of Marin Mersenne (d. 1648), which included mathematicians such as René Descartes (d. 1650), Pierre de Fermat (d. 1665), Jean de Beaugrand (d. 1640), Gilles Personne de Roberval (d. 1675), and Blaise Pascal (d. 1662)

3 Girard Desargues, *Brouillon Projet d'une atteinte aux événements des rencontres du cône avec un plan* (Paris: no publ., 1639). See also Jan P. Hogendijk, 'Desargues' *Brouillon* Project and the *Conics* of Apollonius', *Centaurus* 34 (1) (1991): 1–43.

4 Blaise Pascal, *Œuvres de Blaise Pascal*, 5 vols. (La Haye: Abbé Bossut, 1779); René Taton, 'L'œuvre de Pascal en géométrie projective', *Revue d'histoire des sciences et de leurs applications* 15 (3–4) (1962): 197–252.

5 Ibn al-Haytham, *Kitab al-Manazir*, ed. A. I. Sabra (Kuwait: National Council for Culture, Arts and Letters, 1983 [Books I-III], 2002 [Books IV-V]); Ibn al-Haytham, *The Optics of Ibn al-Haytham, Books I – III, On Direct Vision*, trans. A. I. Sabra (London: Warburg Institute, 1989).

6 Nader El-Bizri, 'In Defence of the Sovereignty of Philosophy: al-Baghdadi's Critique of Ibn al-Haytham's Geometrisation of Place', *Arabic Sciences and Philosophy* 17 (2007): 57–80;

Nader El-Bizri, 'La perception de la profondeur: Alhazen, Berkeley, et Merleau-Ponty', *Oriens-Occidens: Cahiers du centre d'histoire des sciences et des philosophies arabes et médiévales* 5 (2004): 171–184; Nader El-Bizri, 'Le problème de l'espace: approches optique, géométrique et phénoménologique', in *Oggetto e spazio. Fenomenologia dell'oggetto, forma e cosa dai secoli XIII-XIV ai post-cartesiani*, eds. Graziella Federici Vescovini and Orsola Rignani (Firenze: Edizioni del Galluzzo, 2008), pp. 59–70. Refer also to Roshdi Rashed, *Les mathématiques infinitésimales du IXᵉ au XIᵉ siècle*, Vol. IV: *Ibn al-Haytham, méthodes géométriques, transformations ponctuelles et philosophie des mathématiques* (Wimbledon, London: al-Furqan Islamic Heritage Foundation, 2002).

7 Nader El-Bizri, 'Imagination and Architectural Representations', in *From Models to Drawings: Imagination and Representation in Architecture*, eds. Marco Frascari, Jonathan Hale and Bradley Starkey (London: Routledge, 2007), pp. 34–42; Nader El-Bizri, 'Classical Optics and the *Perspectiva* Traditions Leading to the Renaissance', in *Renaissance Theories of Vision*, eds. Charles Carman and John Hendrix (Aldershot: Ashgate, 2010), pp. 11–30; Nader El-Bizri, 'Seeing Reality in Perspective: "The Art of Optics" and the "Science of Painting"', in *The Art of Science: From Perspective Drawing to Quantum Randomness*, eds. Rossella Lupacchini and Annarita Angelini (Dordrecht: Springer, 2014), pp. 25–47.

8 Appolonius of Perga, *Apollonii Pergaei Conicorum libri quatuor* (Bononiae: ex officina A. Benatii, 1566); Appolonius of Perga, *Apollonius de Perge: Coniques*, eds. R. Rashed, M. Decorps-Foulquier and M. Federspiel (Berlin: De Gruyter, 2008–2010).

9 George L. Hersey, *Architecture and Geometry in the Age of the Baroque* (Chicago: University of Chicago Press, 2000), pp. 171–174, 181.

10 If we consider three aligned points A, B, C and three other aligned points a, b, c, the points of intersection of the straight lines (Ab)-(Ba), (Ac)-(Ca), and (Bc)-(Cb) are also aligned.

11 For a given hexagon that is inscribed in a conic section, three pairs of the prolongations of the opposite sides meet on a straight line (posthumously known as 'the Pascal Line').

12 Girard Desargues, *Manière universelle de M. Desargues pour pratiquer la perspective par petit pied, comme le géométral: Ensemble des places et proportions des fortes et faibles touches, teintes ou couleurs* (Paris: Pierre Des Hayes, 1647–1648).

13 Sundararaman Ramanan, 'Projective Geometry', *Resonance* 2 (8) (1997): 87–94.

14 Mechanics enters the liberal arts as set in the *Didascalicon* of Hugues de Saint-Victor, *Libri septem eruditiones didascaliae, Artes mechanicae*: ch.26 (PL 176, col.760): *A Medieval Guide to the Arts*, trans. Jerome Taylor (New York: Columbia University Press, 1961).

15 Girard Desargues, '*Exemple de l'une des manières universelles du S.G.D.L. touchant la pratique de la perspective sans emploier aucun tiers point, de distance ny d'autre nature, qui soit hors du champ de l'ouvrage*' [Manuscript dated on May 1636 and classified in the Bibliothèque Nationale by the code V I, 597] (Paris: Pauillon des Tuylleries, Louvre, 1636). The *folios* correspond with the G. N. Poudra edition of 1864, *op. cit.*, pp. 55–84. See also Kirsti Andersen, 'Desargues' Method of Perspective: Its Mathematical Content, Its Connection to Other Perspective Methods and Its Relation to Desargues' Ideas on Projective Geometry', *Centaurus* 34 (1) (1991): 44–91.

16 Abraham Bosse, *Traité des pratiques géométrales et perspectives enseignées dans l'Académie Royale de la Peinture et Sculpture* (Paris: Bosse Publ., 1665).

17 Carl B. Boyer, 'Robert Grosseteste on the Rainbow', *Osiris* 11 (1954): 247–258; Carl B. Boyer, 'The Theory of the Rainbow: Medieval Triumph and Failure', *Isis* 49 (4) (1958): 378–390; Bruce S. Eastwood, 'Robert Grosseteste's Theory of the Rainbow: A Chapter in the History of Non-experimental Science', *Archives internationales d'histoire des sciences* 19 (1966): 313–322.

18 David C. Lindberg, 'Roger Bacon's Theory of the Rainbow: Progress or Regress?' *Isis* 57 (1966): 235–248.

19 Robert Grosseteste, 'De iride', in *Die Philosophischen Werke des Robert Grosseteste, Bischofs von Lincoln*, ed. Ludwig Baur (Münster i. W.: Aschendorff, 1912), p. 72.

20 Nader El-Bizri, 'Ibn al-Haytham et le problème de la couleur', *Oriens-Occidens: Cahiers du centre d'histoire des sciences et des philosophies arabes et médiévales* 7 (2009): 201–226.

21 Kamal al-Din al-Farisi, *Kitab Tanqih al-manazir* (Hyderabad: Osmania Oriental Publications Bureau, 1928–1929), 2 vols.

22 Theodoricus Teutonicus de Freiburg, *De iride et radialibus impressionibus*, ed. Joseph Würschmidt (Münster i. W.: Aschendorff, 1914).

23 Roshdi Rashed, *Geometry and Dioptrics in Classical Islam* (Wimbledon, London: al-Furqan Islamic Heritage Foundation, 2005), pp. 63, 108, 152, 181; Roshdi Rashed, 'A Pioneer in Anaclastics: Ibn Sahl on Burning Mirrors and Lenses', *Isis* 81 (1990): 464–491.

24 Claudio D'amato, *L'art de la stéréotomie: Les compagnons du devoir et les merveilles de la construction en pierre* (Paris: Librairie du Compagnonnage, 2005).

25 François Derand, *L'architecture des voutes ou l'art des traits et coupe des voutes* (Paris: Sébastien Cramoisy, 1643).

26 Eugène Viollet-le-Duc (d. 1879), *Dictionnaire raisonné de l'architecture française du XIᶜ au XVIᶜ siècle* (Paris: Édition Bance-Morel de, 1854–1868), 'Trait (art du)', Tome 9, pp. 197–214.

27 Girard Desargues, *La manière universelle de Mr. Desargues, Lyonnois, pour poser l'essieu, et placer les heures et autres choses aux cadrans au soleil, par A. Bosse graveur en taille douce, en l'isle du Palais devant la Megisserie, à la Roze rouge* (Paris: Pierre Des-Hayes, 1643). See also the G. N. Poudra edition of 1864, *op. cit.*, p. 366.

28 Amédée-François Frézier, *Éléments de stéréotomie, à l'usage de l'architecture, pour la coupe des pierres* (Paris-Strasbourg: Charles-Antoine Jombert, 1760), 2 vols. [Ré-édition Hachette Livre BNF, 2013]; Amédée-François Frézier, *La théorie et la pratique de la coupe des pierres et des bois pour la construction des voûtes et autres parties . . . ou traité de stéréotomie à l'usage de l'architecture* (Paris-Strasbourg: Charles-Antoine Jombert, 1754–1759), 3 vols. [Ré-édition. Paris: Jacques Laget, 2000].

29 Gaspard Monge, *Géométrie descriptive: Leçons données aux écoles normales* (Paris: Baudouin, 1799).

30 Vitruvius, *On Architecture*, Volume I, Books 1–5, English translation from the Latin by Frank Granger, Loeb Classical Library 251 (Cambridge, MA: Harvard University Press, 1931).

31 Plato, *Platonis Opera*, ed. John Burnet (Oxford: Oxford University Press, 1903); Plato, *The Republic, Books 6–10* (Loeb Classical Library, No. 276) *Greek and English* (Cambridge, MA: Harvard University Press, 1969).

32 Henri-Irénée Marrou, 'Les Arts Libéraux dans l'Antiquité Classique', in *Arts Libéraux et Philosophie au Moyen Âge. Actes du IVᶜ Congrès international de philosophie médiéval, Université de Montréal, 27 août – 2 septembre 1967*, ed. Henri-Irénée Marrou (Paris – Montréal: Vrin – Institut d'Études Médiévales, 1969), pp. 18–19.

33 William Harris Stahl, Richard Johnson, and Evan Laurie Burge, *Martianus Capella and the Seven Liberal Arts* (New York: Columbia University Press, 1971), Vol. 1, p. 156.

34 These were Jean-Victor Poncelet (author of the *Traité des propriétés projectives des figures*. Paris: Bachelier, 1822) and Joseph Diez Gergonne (editor of the *Annales de mathématiques pures et appliquées*. Nimes: La Veuve Belle, 1810–1811). Both became aware of Desargues' works through Michel Chasles, *Traité des coniques* (Paris: Gauthier-Villars, 1865).

35 Desargues' *oeuvres* and Bosse's interpretations were edited in Paris in 1864 by Noël-Germinal Poudra, who was in the French infantry at the rank of *officier supérieur d'état-major* and a graduate of the *École Polytechnique*.

36 This chapter is dedicated *in memoriam* to the historian of mathematics Hélène Bellosta (d. 2011).

Bibliography

Andersen, Kirsti. 'Desargues' Method of Perspective: Its Mathematical Content, Its Connection to Other Perspective Methods and Its Relation to Desargues' Ideas on Projective Geometry', *Centaurus* 34 (1) (1991): 44–91.

Appolonius of Perga. *Apollonii Pergaei Conicorum libri quatuor*. Bononiae: ex officina A. Benatii, 1566.

———. *Apollonius de Perge: Coniques*, eds. R. Rashed, M. Decorps-Foulquier and M. Federspiel. Berlin: De Gruyter, 2008–2010.

Bosse, Abraham. *Traité des pratiques géométrales et perspectives enseignées dans l'Académie Royale de la Peinture et Sculpture*. Paris: Bosse Publ., 1665.

Boyer, Carl B. 'Robert Grosseteste on the Rainbow', *Osiris* 11 (1954): 247–258.

———. 'The Theory of the Rainbow: Medieval Triumph and Failure', *Isis* 49 (4) (1958): 378–390.

Chasles, Michel. *Traité des coniques*. Paris: Gauthier-Villars, 1865.

D'amato, Claudio. *L'art de la stéréotomie: Les compagnons du devoir et les merveilles de la construction en pierre*. Paris: Librairie du Compagnonnage, 2005.

Derand, François. *L'architecture des voutes ou l'art des traits et coupe des voutes*. Paris: Sébastien Cramoisy, 1643.

de Saint-Victor, Hugues. *Didascalicon. Libri septem eruditiones didascaliae, Artes mechanicae*: ch.26 (PL 176, col.760): *A Medieval Guide to the Arts*, trans. Jerome Taylor. New York: Columbia University Press, 1961.

Desargues, Girard. *Exemple de l'une des manières universelles du S. G. D. L. touchant la pratique de la perspective sans employer aucun tiers point, de distance ny d'autre nature, qui soit hors du champ de l'ouvrage*. Paris: Pauillon des Tuylleries, Louvre, 1636.

———. *Brouillon Projet d'une atteinte aux événements des rencontres du cône avec un plan*. Paris: no pub., 1639.

———. *La manière universelle de Mr. Desargues, Lyonnois, pour poser l'essieu, et placer les heures et autres choses aux cadrans au soleil, par A. Bosse graveur en taille douce*. Paris: Pierre Des-Hayes, 1643.

———. *Manière universelle de M. Desargues pour pratiquer la perspective par petit pied, comme le géométral: Ensemble des places et proportions des fortes et faibles touches, teintes ou couleurs*. Paris: Pierre Des Hayes, 1647–1648.

———. *Oeuvres de Desargues*, ed. Noël-Germinal Poudra. Paris: Leiber, 1864.

———. *L'oeuvre mathématique de G. Desargues: Textes publiés et commentés avec une introduction biographique et historique*, ed. René Taton. Paris: Presses Universitaires de France, 1951.

Eastwood, Bruce S. 'Robert Grosseteste's Theory of the Rainbow: A Chapter in the History of Non-Experimental Science', *Archives internationales d'histoire des sciences* 19 (1966): 313–322.

El-Bizri, Nader. 'La perception de la profondeur: Alhazen, Berkeley, et Merleau-Ponty', *Oriens-Occidens: Cahiers du centre d'histoire des sciences et des philosophies arabes et médiévales* 5 (2004): 171–184.

———. 'A Philosophical Perspective on Alhazen's Optics', *Arabic Sciences and Philosophy* 15 (2005): 189–218.

———. 'In Defence of the Sovereignty of Philosophy: al-Baghdadi's Critique of Ibn al-Haytham's Geometrisation of Place', *Arabic Sciences and Philosophy* 17 (2007): 57–80.

———. 'Imagination and Architectural Representations', in *From Models to Drawings: Imagination and Representation in Architecture*, eds. Marco Frascari, Jonathan Hale and Bradley Starkey. London: Routledge, 2007, pp. 34–42.

———. 'Le problème de l'espace: approches optique, géométrique et phénoménologique', in *Oggetto e spazio. Fenomenologia dell'oggetto, forma e cosa dai secoli XIII-XIV ai post-cartesiani*, eds. Graziella Federici Vescovini and Orsola Rignani. Firenze: Edizioni del Galluzzo, 2008, pp. 59–70.

———. 'Ibn al-Haytham et le problème de la couleur', *Oriens-Occidens: Cahiers du centre d'histoire des sciences et des philosophies arabes et médiévales* 7 (2009): 201–226.

———. 'Classical Optics and the *Perspectiva* Traditions Leading to the Renaissance', in *Renaissance Theories of Vision*, eds. Charles Carman and John Hendrix. Aldershot: Ashgate, 2010, pp. 11–30.

———. 'Seeing Reality in Perspective: "The Art of Optics" and the "Science of Painting"', in *The Art of Science: From Perspective Drawing to Quantum Randomness*, eds. Rossella Lupacchini and Annarita Angelini. Dordrecht: Springer, 2014, pp. 25–47.

al-Farisi, Kamal al-Din. *Kitab Tanqih al-manazir.* Hyderabad: Osmania Oriental Publications Bureau, 1928–1929, 2 vols.

Field, Judith V. and Gray, J. J. *The Geometrical Work of Girard Desargues.* Dordrecht-Berlin: Springer, 1987.

Frézier, Amédée-François. *La théorie et la pratique de la coupe des pierres et des bois pour la construction des voûtes et autres parties . . . ou traité de stéréotomie à l'usage de l'architecture.* Paris-Strasbourg: Charles-Antoine Jombert, Paris, 1754–1759, 3 vols. [Ré-édition. Paris: Jacques Laget, 2000].

———. *Éléments de stéréotomie, à l'usage de l'architecture, pour la coupe des pierres.* Paris-Strasbourg: Charles-Antoine Jombert, Paris, 1760, 2 vols. [Ré-édition Hachette Livre BNF, 2013].

Gergonne, Joseph Diez Gergonne. *Annales de mathématiques pures et appliquées.* Nimes: La Veuve Belle, 1810–1811.

Grosseteste, Robert. *Die Philosophischen Werke des Robert Grosseteste, Bischofs von Lincoln,* ed. Ludwig Bauer. Münster i. W.: Aschendorff, 1912.

Hersey, George L. *Architecture and Geometry in the Age of the Baroque.* Chicago: University of Chicago Press, 2000.

Hogendijk, Jan P. 'Desargues' *Brouillon* Project and the *Conics* of Apollonius', *Centaurus* 34 (1) (1991): 1–43.

Ibn al-Haytham. *Kitab al-Manazir,* ed. A. I. Sabra. Kuwait: National Council for Culture, Arts and Letters, 1983, 2002.

———. *The Optics of Ibn al-Haytham, Books I – III, On Direct Vision,* trans. A. I. Sabra. London: Warburg Institute, 1989.

Lindberg, David C. 'Roger Bacon's Theory of the Rainbow: Progress or Regress?' *Isis* 57 (1966): 235–248.

Marrou, Henri-Irénée. 'Les Arts Libéraux dans l'Antiquité Classique', in *Arts Libéraux et Philosophie au Moyen Âge,* ed. Henri-Irénée Marrou. Paris – Montréal: Vrin – Institut d'Études Médiévales, 1969. pp. 6–27.

Monge, Gaspard. *Géométrie descriptive: Leçons données aux écoles normales l'an 3 de la république.* Paris: Baudouin, *imprimeur du corps législatif et de l'institut national,* 1799.

Pascal, Blaise. *Œuvres de Blaise Pascal.* La Haye: Abbé Bossut, 1779, 5 vols.

Plato. *Platonis Opera,* ed. John Burnet. Oxford: Oxford University Press, 1903.

———. *The Republic, Books 6–10* (Loeb Classical Library, No. 276) *Greek and English.* Cambridge, MA: Harvard University Press, 1969.

Poncelet, Jean-Victor. *Traité des propriétés projectives des figures.* Paris: Bachelier, 1822.

Ramanan, Sundararaman. 'Projective Geometry', *Resonance* 2 (8) (1997): 87–94.

Rashed, Roshdi. 'A Pioneer in Anaclastics: Ibn Sahl on Burning Mirrors and Lenses', *Isis* 81 (1990): 464–491.

———. *Les mathématiques infinitésimales du IXe au XIe siècle,* Vol. IV: *Ibn al-Haytham, méthodes géométriques, transformations ponctuelles et philosophie des mathématiques.* Wimbledon, London: al-Furqan Islamic Heritage Foundation, 2002.

———. *Geometry and Dioptrics in Classical Islam.* Wimbledon, London: al-Furqan Islamic Heritage Foundation, 2005.

Stahl, William Harris, Johnson, Richard, and Burge, Evan Laurie. *Martianus Capella and the Seven Liberal Arts.* New York: Columbia University Press, 1971.

Taton, René. 'L'œuvre de Pascal en géométrie projective', *Revue d'histoire des sciences et de leurs applications* 15 (3–4) (1962): 197–252.

Teutonicus de Freiburg, Theodoricus. *De iride et radialibus impressionibus,* ed. Joseph Würschmidt. Münster i. W.: Aschendorff, 1914.

Viollet-le-Duc, Eugène. *Dictionnaire raisonné de l'architecture française du XIe au XVIe siècle.* Paris: Édition Bance-Morel de, 1854–1868.

Vitruvius. *On Architecture,* Volume I, Books 1–5, english translation from the Latin by Frank Granger (Loeb Classical Library 251). Cambridge, MA: Harvard University Press, 1931.

3 Galileo's limit

Mechanical sciences' technologies of sight and the translation of analogical representations into diagrammatic illustrations

Federica Goffi

The historical study of graphic notations and representations related to the mechanical sciences during the sixteenth to seventeenth centuries shows the development of new technologies of vision that evolved out of merging physiological knowledge and Galilean mechanics. The visual language associated with a newly interpreted Vitruvian *firmitas* is nurtured by Renaissance humanistic culture with a new impulse offered by anatomical studies. A study of the evolution of the provisional frames used in the narrative development of the *firmitas* in the sixteenth to seventeenth centuries is essential in understanding the translations taking place from anthropomorphic architecture to mechanical engineering.

With Galileo Galilei's *Dialogues Concerning Two New Sciences* (1638), both geometry and materials lost cultural and anthropomorphic character. The late Renaissance scientist cast doubts on the theory of proportions. He substituted experiential with experimental knowledge to achieve certainty through an attempt to calculate invisible forces acting within building elements. A predictive assessment replaced a projective one. The questionable separation of the *firmitas* from the Vitruvian triad opened a challenging gap between architectural modes of conceiving and those of the mechanical sciences, and it reduced the art of construction to structural optimization and economy.[1] Firmness, as Emmons explains it, is not just a measurable condition but also, and more importantly, a human experience and a cultural condition.[2] Furthermore, Vitruvian *firmitas* was interpreted in the early Italian Renaissance architectural treatise not merely as a concern with solidity per se, but rather included concepts of solidity as a means to achieve endless duration as the hereafter (Lat. *Sempiternitas*).[3]

A micro-historical close up look into the technologies of vision adopted,[4] such as sciography, axonometric and vertical sections, to represent architectural details and studies of human anatomy in the works of Leonardo da Vinci, and later Galileo Galilei and Giovanni Alfonso Borelli, allows us to appreciate a telling transformation from analogical representations to diagrammatic illustrations. This places emphasis on the crucial role played by the annotative drawing techniques, used to develop theories of structures based on the body analogy, in the transformation of the technologies of vision that shape the architectures of sight. Visual references to the body were removed. The visual imagery used in the fabrication and transportation of meaning in the emerging mechanical science oriented the production of so-called *scientific* information through increased levels of abstraction based on descriptive geometry and Cartesian dualism.

Latour defines the "immutable mobile" as the essence of an idea carried over from one image to another.[5] In this act of mediating scientific arguments through progressive substitutions, the visual medium is not a neutral device. When taken singularly, an image is but a provisional element, yet it affects the translation of meaning occurring in a sequence of elaborations. Consequently, with the production of each new image, what is carried over is just as important as what is left behind.

Latour introduced the terminology of 'provisional frame' to refer to an image which articulates meaning visually and is part of a larger scientific or artistic discourse. While referring to the overarching significance of his argument, a dual terminology of 'provisional images' and 'fixed frames' is used to indicate, respectively, analogical drawings that transform through time and diagrammatic illustrations devoid of materiality and scale, emphasizing changes occurred in mechanical science with the search for exact knowledge based on geometry and mathematical calculations.[6]

The traditional empathic reading implicit in Western Renaissance architecture suggesting a meaningful use of the anthropomorphic analogy in the reading of proportions as pleasing to the eye-brain and the empathic reading of tensional states through our postural sense and embodied musculoskeletal experience become negligible.[7] Structural empathy allows us to intuitively understand building elements by mimetically experiencing tensional states through the act of looking at images and drawings, which act as mirrors of the body.[8] Image after image, and frame after image, the transformation draws attention away from bodily references and towards abstracted geometry and theorized forces.

A paradigm shift in tectonic imagination begins with Galileo's concepts of structural optimization. This approach translated into a belief in historical improvement in the use of materials. During the following century, the transformation from architectural imagination to problem solving is slow but progressive, leading to a separation of competencies and methods of work between the emerging engineering science and architectural design; while the former relied on measurements and repeatable experiments, the latter was guided primarily by the anthropomorphic analogy and experience. In pre-Galilean science, the understanding of stability was conjoined with the overall conception of a building's fabric conjured as a mirror of cosmic order.[9] The proportionality of the body reflected the harmonic union of firmness, commodity and delight *(firmitas, utilitas, venustas)* understood as an inseparable Vitruvian triad.[10]

In addition to the optical theories of Galileo influencing Western thought and architecture, so did his related ideas on structures and the body. Crucially, both relied on the visioning techniques of drawing to achieve their status and have the effect they ultimately would. These arguments have a continued historical significance and are critical to an architect's tectonic imagination merging material transformations and culture. The power of a technology of vision employing bodily constructions in section and sciography is that through them we empathize with structures. With the removal of the body, the underlying assumption is that the structure of a building, and the relation between layers that constitute an integral whole, allowing a reading of anatomy and poetry through the skin as a register of depth and culture, is seldom questioned, reducing architecture to a skin-deep practice.

Galilean *firmitas*

The modern beginning of mechanical science traces back to Galilei's *Dialogues Concerning Two New Sciences* (1638). This work marks a paradigm shift that separated the methods to conceive the structure from the overall architecture. According to

Galileo, a scientific investigation of the *firmitas* had yet to be developed.[11] Vitruvius' *Ten Books on Architecture* had a prominent position as the recovered link with antiquity.[12] Galileo intended to lay the foundations for a new science arguing for a novel reading of the book of nature through empirical observations. The separation of the *firmitas* from the Vitruvian triad finds an *incipit* in the critique of the theory of proportions.[13]

Renaissance architects expand on the Vitruvian theory. Vitruvius warns about scaling machines in Book X.[14] Filarete discusses proportions as a matter of proper relations between parts rather than as a scaling device. He acknowledges the variation of human beings' size and distinguishes the very large (giant) and very small (dwarf) as counterfeits that deviate from nature's work.[15] In this way, the idea of proportion was known to hold true across a given range related to the ratio of the work.[16]

Galileo owned a copy of Palladio's *I Quattro Libri dell'Architettura* (1570).[17] Palladio's inventions for timber truss bridges were based on proportionality. For one of his designs, he explained that the height of the bridge's posts should be equal to the width of the river divided by eleven; this implied that proportions hold true regardless of size.[18] Galileo counters this position stating, "one cannot argue from the small to the large, because many devices which succeed on a small scale do not work on a large scale."[19]

Sebastiano Serlio's *Regole Generali dell'Architettura* (1537) and Giovanni Antonio Rusconi's *Della Architettura* (1590) provided the Vitruvian text with a body of images re-actualizing the architectural principles.[20] Galileo's text introduced theoretical abstraction with the hypothesis of a homogeneous material.[21] His work builds on sixteenth-century anticipations of mechanical problems by Leonardo da Vinci, Niccolò Tartaglia, Giovanni Battista Benedetti, Guido Ubaldo and the Hellenistic mechanics by Aristotle and Archimedes.[22] Leonardo's sketches of beams, pillars, trusses and arches (Madrid Codex 8937, Milan 1490) linked physiology and mechanics projecting central questions.

During this period, the pseudo-Aristotelian *Questions of Mechanics* supplemented the theory of proportions with observations concerning mechanical physiology.[23] Aristotelian mechanics was a mixed science uniting mathematics and physics. It investigated physiological mechanics through analogy with simple machines such as lever and balance. Questions such as "why is it that when people rise from a sitting position, they always do so by making an acute angle between the thigh and the lower leg and between the chest and the thigh, otherwise they cannot rise?"[24] were reexamined in Borelli's '*De Motu Animalium*' (1680). Borelli studied at the Accademia del Cimento with Galileo's disciples. His treatise focused on the muscles' leverage action to explain the mechanics of the body.

Pseudo-Aristotelian philosophies conceived mechanics as artful resourcefulness superseding nature. Following the pseudo-Aristotelian tradition, Galileo sought to explain mechanical problems by means of the lever. However, he affirmed that the mastery of mechanical skills arises from an understanding of the very principles that govern nature.

Before studying mathematics, mechanics and astronomy under Ostilio Ricci, Galileo studied medicine at the University of Pisa (1581–1585). His theories were born within a tradition of architectural anthropomorphism and humanist culture, where both architects and scientist practiced the anatomical table.[25] The skeleton and muscles systems were paradigmatic models reciprocal to a building's structure.[26]

The eyes of an idiot perceive little by beholding the external appearance of the human body, as compared with the wonderful contrivances which a careful and practiced anatomist or philosopher discover in that same body when he seeks out the use of all those muscles, tendons, nerves and bones.[27]

Galileo contemplates a flayed architecture. The body is the paradigmatic model used to revise mechanics, yet shifting from an appreciation of visible effects (proportions) to the invisible ones (forces). He draws attention on building elements rather than the body, which is visually and graphically omitted from representation, with the exception of one image chosen to set the limit to the theory of proportions.[28]

A contextual analysis of the images found in *Two New Sciences* within the broader context of sixteenth-century visual culture reveals that analogical representations transformed into diagrammatic illustrations. The fabrication and transportation of meaning happened through substitution of provisional images lead through increased levels of abstraction to the production of so-called *scientific* information. The transformation was achieved by obscuring the presence of the body as the original medium for the intuitive understanding of structures.

Formerly, a building's fabric was appreciated through knowledge of materials, geometry, measurements and experience. Orthographic projections allowed the appreciation of proportions. Observing buildings in a state of disrepair manifested the ravaging effects of time. The pre-Galilean art of edification allowed knowledge crossover from one structure to another, one building to another and, importantly, from one builder to another. Payne explains that Gallacini's organological understanding of mechanics was based on a mediation between experience and (philosophical) abstraction.[29] Despite his fascination with the know-how displayed at the Venetian Arsenal by shipbuilders, Galileo moves away from experiential knowledge towards experimental knowledge based on repeatable and measurable phenomena.[30]

Galileo's science: From analogical representations to diagrammatic illustrations

A close-up analysis of Galileo's provisional images allows for the understanding of his vision of mechanics, and its derivation from physiology, but also how he distanced himself from it. Pre-Galilean *firmitas* relied on a mimetic process with nature, which held the human body as the paradigmatic mirror of cosmic order. Provisional images relied on visual analogy to weave their understanding of the body through tectonic elements.

Galileo linked mechanical skills and architecture through the study of details.[31] A cantilevered beam is used to postulate the so-called problem of maximal dimensions (Fig. 3.1), while a suspended truncated column is used to study tension (Fig. 3.2). The technology of vision adopted is *sciografia*: a false perspective used in sixteenth-century architectural treatises. Serlio's drawings of ancient Roman buildings (Book III) provide an example where ruins manifest anatomical assembly and fabrication.[32]

Scolari explains that by the early sixteenth century, the way was paved for parallel projections to become "the accredited method of representation for engineering science."[33] Galileo could have used a true parallel projection but chose sciography instead for a majority of details. Through the straight drawing of one face, *sciografia* offered a didactic sense of assembly, geometry, measurability and materiality while enhancing three-dimensional qualities by outlining the receding sides into a vanishing point.[34]

Figure 3.1 Cantilevered Beam (1638), Galileo Galilei. Istituto e Museo di Storia della Scienza, Florence, Rare collection 075. © Courtesy of the Museo Galileo, Florence, Library

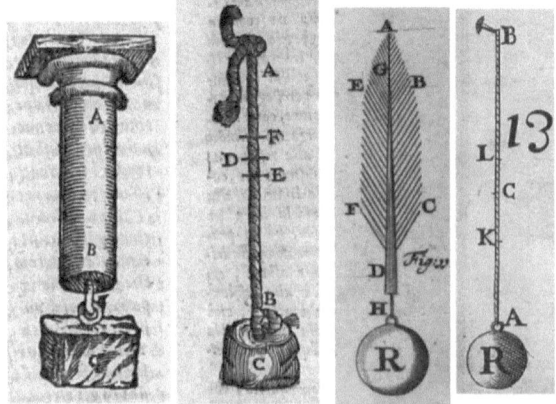

Figure 3.2 Left two images: (1638), Galileo Galilei. Istituto e Museo di Storia della Scienza, Florence, Rare collection 075. © Courtesy of the Museo Galileo, Florence, Library. Right two images: 1680, Giovanni Alfonso Borelli. *De Motu Animalium*. Public Domain

However, the images accompanying his work are not to scale, nor are they rigorously well drawn with compass and ruler, thus demoting representation to illustration. Lacking a graphic scale in the work, he could have just as well omitted images all together. Their presence might indicate his reliance on mechanical devices rather than on continued observation of actual architectural details.[35]

Galileo's decision to illustrate mechanical questions by means of ruins is also significant. Sciography was the technology of vision chosen by Rusconi to demonstrate details assembly by peeling away layers.[36] Serlio stated that the greatness of Rome could be seen in its ruins and adopted sciography as a suitable technology of vision to demonstrate assembly.[37] Rusconi's theater of ruins is like the theater of the human body, exposing an anatomized fabric. Anatomical lexicon finds its way into the treatise, constructing verbal and visual analogies.

Galileo is determined to set the limits of the theory of proportions. It is perhaps for this reason that he avoids using images of the body, with the sole exception of what is presumably a thigh bone, the largest and thickest bone in the skeleton. The thigh bone symbolizes human anatomy and directs attention to nature's own ratio in the work: a self-chosen limit. Galileo introduced his interrogation of the theory of proportions by discussing the problem of maximum dimensions and affirmed,

> *If one wishes to maintain in a great giant the same proportion of limb as that found in an ordinary man he must either find a harder and stronger material for making the bones, or he must admit a diminution of strength in comparison with man of medium stature; for if his height be increased inordinately he will fall under his own weight.*[38]

According to Galileo, material imperfections do not account for this observation. He proceeds to study a suspended, truncated column from which a roughly shaped stone is hung (Fig. 3.2). In another provisional image, he substitutes the column with a rope. The rope visualizes the hypothetical concentration of tensile forces along a vertical axis. He demonstrated that the resistance of a prism of constant thickness subjected to uniform tension is independent from the length of the prism.

Moving from one provisional image to the next to arrive at stable frames and exact knowledge lead through a process of geometric abstraction to removing all references to the human body. A straight line without thickness along which forces are concentrated is used to conceptualize a solid body subject to tension, completing the transformation from analogical representations to diagrammatic illustrations.

Provisional frames before and after Galilean science: Leonardo, Teofilo Gallacini and Alfonso Borelli

Leonardo's studies of muscular action anticipated Galileo's understanding of tensile stress and Borelli's studies of bodily movement. Sections and sciography drawings were used to demonstrate his knowledge of anatomy and layered superimposition of muscles working in synergy with the skeleton.

> *You will never avoid confusion in the representation of muscles and of their position, origin and insertion if you do not make first a representation of the slender using threads, and thus you will be able to represent one over the other as Nature has placed them' explaining the systems which enable the human arm to be turned, and the mechanism of the shoulder and ankle.*[39]

Leonardo's analogical drawings substituted ropes in place of muscles conveying through a visual metaphor that every bundle of muscle fiber uses its power along the line of its length (Fig. 3.3). Leonardo's drawings are based on meaningful experiences *(esperienza sensata)* and corporeal sensing.[40] The body is the paradigmatic model for

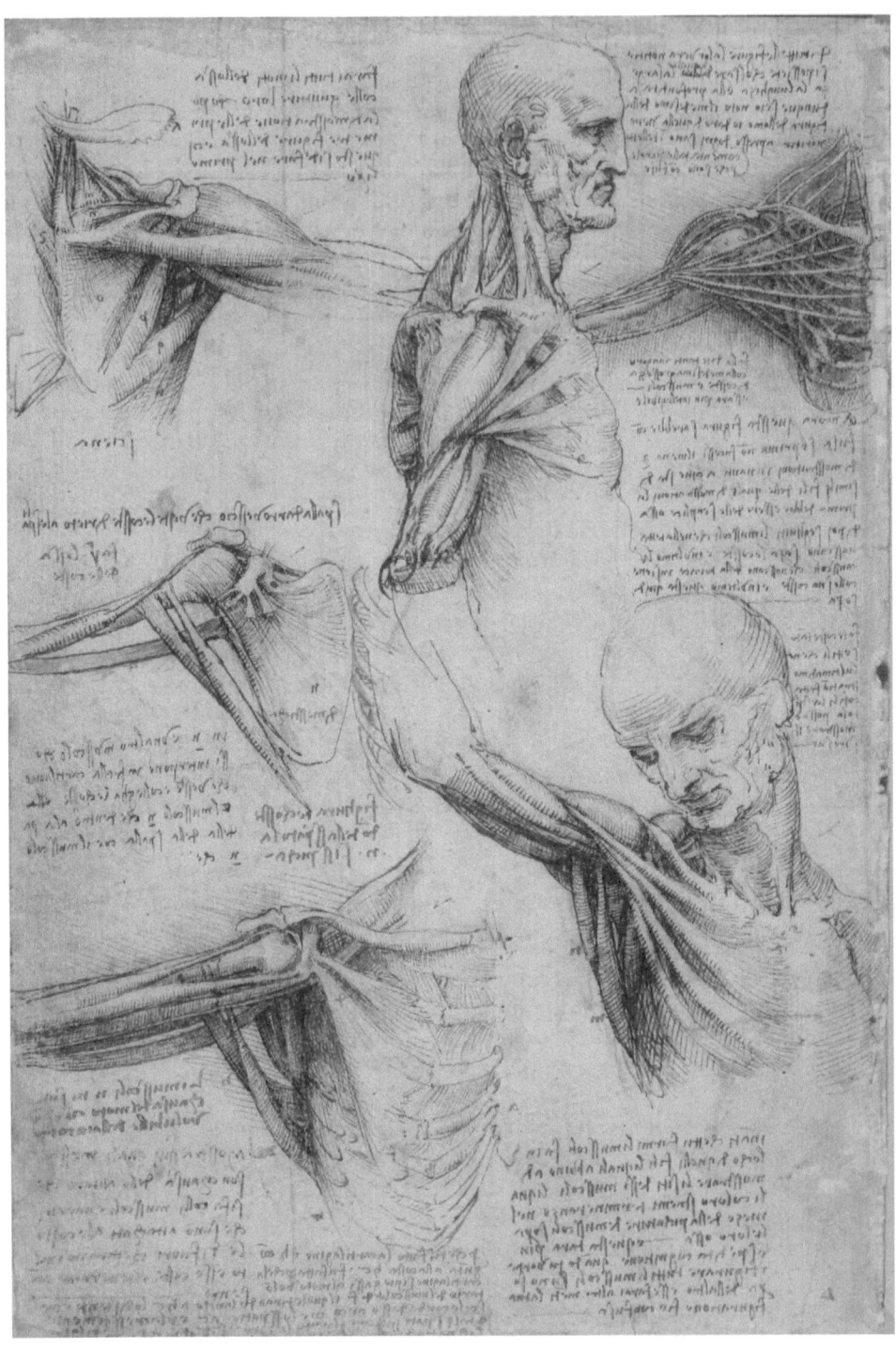

Figure 3.3 Anatomical study of the shoulder (ca. 1510–1511), Leonardo Da Vinci. Royal Collection Trust (RCIN 919003) / Her Majesty Queen Elizabeth II 2015

the understanding of both physiology and mechanics. Forces were, according to him, spiritual virtues expressing invisible powers infused in the body.[41]

Borelli chose to represent the *"muscolorum vera figura"* (real figure of a muscle) as a vertical filament from which fibers depart obliquely in parallel lines.[42] In another provisional image, a muscle is substituted by a vertical rope suspending a weight (Fig. 3.2). This bio-mechanical device reminds us of Galileo's explorations of tension in a suspended column, which "would break like a rope".[43]

Borelli explained the bio-mechanics of the humerus-ulna cantilever system as a leverage (Fig. 3.4). This bodily configuration is treated analogously to Galileo's own cantilever beam (Fig. 3.1). Galileo explained that the beam is one of the arms of the

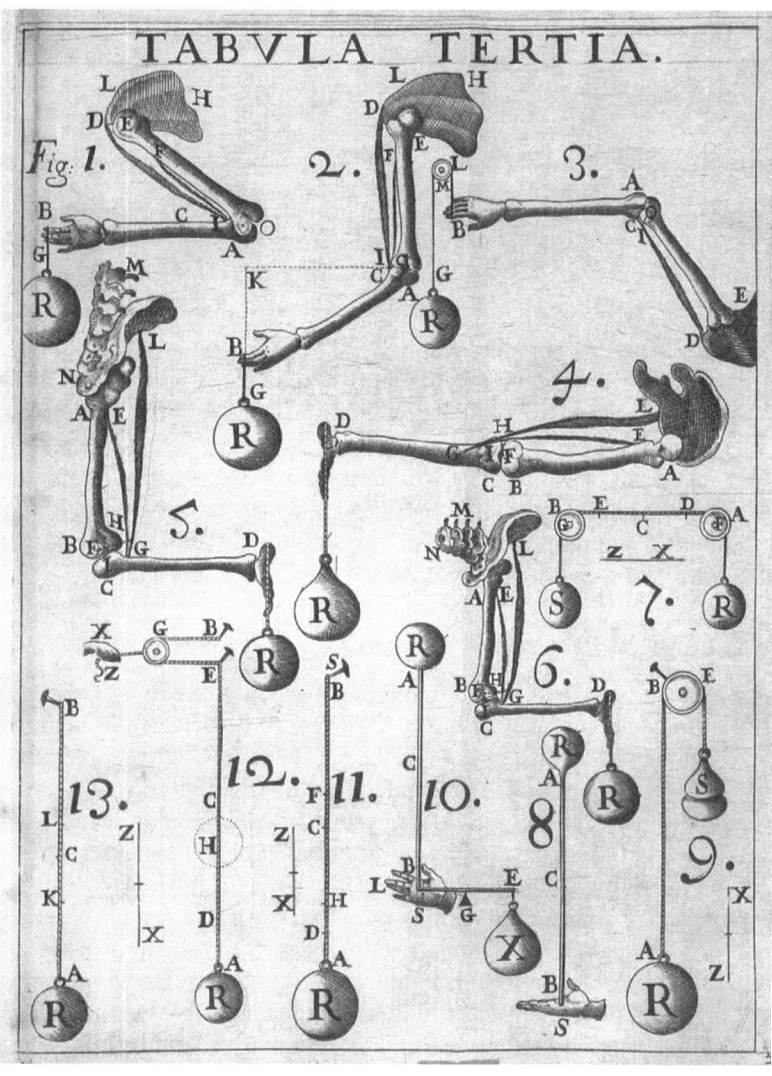

Figure 3.4 De Motu Animalium (1680: Table III), Giovanni Alfonso Borelli. © Public Domain

Figure 3.5 Study of the strength of the muscles of the upper limb. Ms. H (IFP) cc 43v and 44r (1493–94), Leonardo da Vinci. Bibliothèque de l'Institut de France, Paris

lever, the hinge is at the joint between beam and wall and the beam's vertical connection into the wall (AB) is the other arm.[44]

Leonardo's drawings of upper limbs lifting a weight show pulleys and ropes raising weights (Fig. 3.5). By comparison, the conflation of body and machine in Leonardo's and Borelli's work is absent in Galileo's illustrations, where the body is a visibly omitted figure. Galileo was aware of the works of his predecessors and contemporaries, including those of Teofilo Gallacini (1564–1641), who drew out-from-within the body invisible geometric diagrams. Payne argued that he is the first to separate ornament from structure, *ingegno* from science, through "enlivened mechanics."[45]

Descartes and the abstraction of geometry

Descartes was aware of Galileo's work, but they never met. In 1638, Descartes replied to Father Marin Mersenne who solicited a review of *Two New Sciences*, offering a confidential evaluation and stating that he found the work digressive, much like a "building without foundations".[46] He agreed on the method of investigation; however, according to him, Galileo omitted a necessary discussion of primary causes. He discredits the foundation of a new science considering the self-evidence of the problem of maximum dimensions and the lack of proof in some instances.

A few years later, Descartes is in the position of having to defend his own *Discourse on Method* (1637) against objections. He theorized the dualism of body and mind, while also admitting their commingled nature. He was determined to demonstrate the theoretical existence of geometrical and mathematical truths separate from a sensible body. According to him, a "surface [. . .] is merely a mode and hence cannot be part of a body; for a body is a substance, and a mode cannot be a part of a substance."[47] Cartesian geometry is thin to the point of lacking any depth.

Curiously, when Descartes defended the theorized dichotomy of mind and body, he relied on a metaphor. He explained that this dichotomy is analogous to the relation between heaviness – that is a property of a body – and the sensible body: "heaviness is diffused throughout the entire body that is heavy, still I did not ascribe to it that very extension which constitutes the nature of the body".[48] This duality is further elaborated through a visual analogy describing a device similar to one studied by Galileo:

> *heaviness could be contracted to a mathematical point. Indeed, I saw that heaviness, while remaining coextensive with the body, which is heavy, could exert its entire force in any part of the body. For if the body were suspended by a rope, it could pull on the rope with all its force just as if this heaviness were only in the part touching the rope and were not diffused throughout the remaining part – regardless of the part to which the rope might be attached. And it is precisely this way that surely I now understand the mind to be coextensive with the body, the whole of the mind in the whole of the body. And the whole of the mind in every one of its parts.*[49]

In the same text, heaviness is discussed in relation to gravity: "I believed that heaviness carried bodies towards the center of Earth as if it contained within itself some knowledge."[50] A drawn visual metaphor is found in William Hodson's 1640 frontispiece of *The Divine Cosmographer, or a Brief Survey of the Whole World* (Fig. 3.6). Hodson stated that Earth is firmly suspended by a cord held in God's own hand, emerging from clouds in a sky illuminated by the symbol of the trinity.[51]

Simon Stevin defines in *Les Oeuvres Mathematiques* (1634: 434) the center of gravity of a globe and states that this is a symmetrical homogeneous weighty body, where the geometric center and the center of gravity coincide. He depicts the Hand of God emerging from the clouds, while holding a cord stretched towards the center of balance of the geometrical figure (Fig. 3.7).

Drawn a few years later, Galileo's depiction of a rope suspending a weight reminds us of Stevin's device. The Galilean suspended column reveals much more than what it aimed to illustrate. It is an index of the intimate nature of the machine. When considered within the context of treatises of the period, one can now view the column as a provisional image, concealing a continued interest in matters of cosmography.

In Galileo's provisional frames, the column and the rope are not suspended by God's hand. His lever systems, unlike those of Stevin balanced by God's hand, firmly rest on earth. Galileo wrote the new science while under house arrest, after his trial and sentence in 1633. He moved his interests away from matters that could attract attention from the Church of Rome, such as the relation of mind and body, or the relation of God with the world, or matters of gravity and magnetism. Perhaps the lack of foundations lamented by Descartes had to do with this concern. The publication of Galileo's new work had to be secured in a non-Catholic country because of the veto against publishing his work.

Galileo's suspended column leaves one to wonder if in fact the analogical frames did not conceal other questions relating to the nature of gravity. In his studies on magnetism (1600–1609), he made use of several armed loadstones.[52] Benedetto Castelli, a student of Galileo, describes in his *Discourse on the Loadstone* (1639–40) a device (Fig. 3.8) presented by Galileo to Ferdinand II (1608 ca.). Castelli states that Earth, which is referred to as the "great loadstone", shares a similar nature with loadstones.

Figure 3.6 *The Divine Cosmographer, or a Brief Survey of the Whole World,* Frontispiece (1640), William Hodson. The Folger Shakespeare Library (STC 13554), Washington, DC

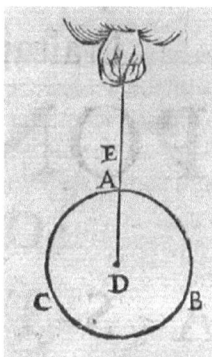

Figure 3.7 *Les Oeuvres Mathematiques'* [. . .] (1634: 434). Simon Stevin. The Folger Shakespeare Library (Q155 S77), Washington, DC

Figure 3.8 Wood post, loadstone, brass hanger, iron sepulcher, wood (1608 ca.) Museo Galileo (Inv. N. 2431) Florence. Photo by Franca Principe

The armed loadstone is analogically related to the devise used to study tensional states in the mechanics.[53] They share a comparable immutable mobile to investigate tensional states induced, respectively, by a magnet in a rope when lifting the cover of a miniaturized sepulcher and the tensional state induced by a stone in a rope, which is pulled down by gravity. Johannes Kepler (*Astronomia Nova*, 1609) was convinced that a connection existed between gravitational phenomenon, planetary motion and magnetism.

Galileo's preoccupation with mechanical science might have veiled other scientific queries. Notably, according to Vitruvius, "all machinery is derived from nature, and is founded on the teaching and instruction of the revolution of the firmament."[54] From this standpoint one can see how the Galilean *firmitas* might hold relations to the mysteries of the *firmamentum*. However, in the process of concealing such relations, the link was obscured and perhaps severed.

Conclusions: Iconoclastic attitudes towards body imagery in mechanical science

Galileo was immersed in a cultural milieu that deemed bodily visual representation essential to communicate meaning, yet he turned to mathematical calculation and diagrammatic representation to transform that. His schematization of details introduced abstracted fixed frames, establishing the inherent limit of Galileo's own conceptions. The emerging mechanical science transformed image after image, frame after image, from analogical representations to diagrammatic illustrations, ruling out the reliance on the theory of proportions. The human body, understood as the paradigmatic instrument upon which to model man-made fabrics became a concealed element.[55]

Leonardo's and Rusconi's hybrid fabrics overlaying the human body and architecture suggest that they formed an entangled and reciprocal system of knowledge. Joseph Rykwert explained that the word *fabrica* crossed over fields such as cosmography, anatomy and architecture, and it was used to indicate in a "constant metaphoric chain" the fabric of the heavens, the fabric body and the architecture's own fabric.[56]

Despite the increased levels of abstraction achieved by graphic engineering, the body remains a fundamental hidden paradigm. The shift that took place entailed moving from an understanding of the fabric of the body, to the body as machine.[57] The translation of provisional images reached a point where the transformation from analogical to diagrammatic representations is complete and the body, understood as point of departure, is lost, both visually and culturally. Yet the analogical use of the body as design instrument is essential to a rejoined body of architecture and engineering science. The way in which we understand the world is based on the fact that the body is present in spatio-temporal continuity with the world. Technologies of vision relying on analogical representations are an effective tool to imagine and understand man-made structures.

Objective thinking has ignored the body because it was believed to introduce subjective elements.[58] The iconoclasm of the body allowed for developing mechanical science through increased levels of abstraction. Analogical representations convey an empathetic understanding of structures that allows us to intuitively comprehend the essence of engineering problems. During the sixteenth to seventeenth centuries, architectural treatises, mechanical science, bio-mechanics, geometric and cosmological treatises shared provisional images for the transportation of meaning, thus demonstrating the commingled nature of the inquiries.[59]

At the time of Galileo's own writings, materials such as steel, which approximates the ideal characteristic of homogeneity postulated in mechanics, were non-existent. Modern engineering science does not apply well to historic materials. Safety factors in solid timber are greater than in steel, reducing its use to a risk level that is considered acceptable. Solid timber is seldom used in contemporary construction, despite being one of the most widely used historical materials. The new science was based on a series of hypotheses and levels of abstraction that in essence also defined new limits.

Notwithstanding the occasional anomaly – such as the work of contemporary engineer and architect Santiago Calatrava, in which he intuitively sketches bodily postures to reflect on complex structural systems – the body analogy is for the most part shunned from engineering science.[60] Yet it is argued to be a powerful means of empathizing with structures by offering through the technologies of vision an intuitive and imaginative means to design. The architectural imagination could rely on multiple systems of knowledge including, but not limited to, mathematical calculations.

Overly fixed frames contribute to stalling the imagination and flattening the architect's role to fashioning visual impressions. Fixed frames defining engineering science today lack a physical body referent. The obscured body is a latent ghost in the machine awaiting imaginative re-incorporation through contemporary technologies of vision. To initiate a shift from the application of iconoclastic disembodied frames to new iconophilic conceptions requires a new discourse, which may arise from the awareness of the trans-historical re-imagination of provisional images through the technologies of sight.

Notes

1 Alberto Pérez-Gómez, *Architecture and the Crisis of Modern Science* (Cambridge, MA: MIT Press, 1983), 25, 39–40, 166–167, 237–267.
2 Paul Emmons, "A Window to the Soul: Depth in Early Modern Section Drawing," in *Architecture Post Mortem: The Diastolic Architecture of Decline, Dystopia, and Death*, ed. Donald Kunze, David Bertolini and Simone Brott (Surrey, UK: Ashgate, 2013), 163–166.
3 Francesco di Giorgio Martini, *Trattato di Architettura, Ingegneria e Arte Militare* (Milano: Il Polifilo, 1967 [1474–1482]), 370. Federica Goffi, *Time Matter(s): Invention and Re-Imagination in Built Conservation: The Unifinished Drawing and Building of St. Peter's in the Vatican* (Surrey, UK: Ashgate, 2013), 154–155.
4 Carlo Ginzburg, "Microhistory: Two or Three Things That I Know about It," in *Critical Inquiry*, 20, 1 (1993): 10–35.
5 Bruno Latour, "How to be Iconophilic in Art, Science and Religion?" in *Picturing Science, Producing Art*, ed. Caroline A. Jones and Peter Galison (London: Routledge, 1998), 421–422.
6 Henk Bos, *Redefining Geometrical Exactness. Descartes' Transformation of the Early Modern Concept of Construction* (New York: Springer, 2001), 231–254.
7 Harry Francis Mallgrave, *The Architect's Brain: Neuroscience, Creativity, and Architecture* (West Sussex: Wiley-Blackwell, 2011), 1–84.
8 One notable exception within the history of engineering science is the case of Henry Fowler and Benjamin Baker's design for the Forth Railway Bridge (Scotland 1882–90). Vilayanur S. Ramachandran and Sandra Blakeslee, *Phantoms in the Brain: Probing the Mysteries of the Human Mind* (New York: William Morrow & Company, 1998).
9 Antonio di Pietro Averlino Filarete, *Trattato di Architettura* (Milano: Il Polifilo, 1972), I, 14.
10 Vitruvius, *The Ten Books on Architecture* (New York: Dover Publications, 1960), 17. Leon Battista Alberti, *On the Art of Building in Ten Books*, ed. Joseph Rykwert and Neil Leach (Cambridge, MA: MIT Press, 1991), 9, 25, 426–427.
11 Vitruvius, *The Ten Books*, 16–17.

12 Vaughan Hart and Peter Hicks, *Paper Palaces: The Rise of the Renaissance Architectural Treatise* (New Haven: Yale University Press, 1998), 6.
13 Vitruvius, *The Ten Books*, 103.
14 Vitruvius, *The Ten Books*, 316. Vitruvius, *De Architettura* (Milano: Einaudi, 1997), 1358–1359.
15 Filarete, *Trattato di Architettura*, 14–15.
16 Paul Emmons, "Size Matters: Virtual Scale and Bodily Imagination in Architectural Drawing." *Architecture Research Quarterly* 9, 2–3 (2005): 227.
17 Galileo's personal book collection was compiled by Favaro (Museo Galileo, Florence).
18 Andrea Palladio, *The Four Books of Architecture* (New York: Dover Publications, 1965 [1570]), Book III, Plate V. Emmons, "Size Matters," 228.
19 Galileo Galilei, *Dialogues Concerning Two New Sciences*, trans. Alfonso de Salvio and Henry Crew (New York: Dover Publications, 1954 [1638]), 2.
20 Daniele Barbaro, *Vitruvio. I Dieci Libri dell'Architettura* (Milano: Il Polifilo, 1987 [1567]). Ceasare Cesariano, *De Architectura* (London: Benjamin Blom, 1968 [1521]).
21 Galielo, *Dialogues Concerning Two New Sciences*, 2–3. Material testing begins with the work of Giovanni Poleni (1683–1761) and Pieter van Musschenbroek (1692–1761).
22 Edoardo Benvenuto, *An Introduction to the History of Structural Mechanics: Part I: Statics and Resistance of Solids* (New York: Springer-Verlag, 1991).
23 Galileo owned Leonico translated edition of *Conversio Mechanicarum Questionum Aristotelis cum Figuris, et Annotationibus quibusdam* (1525). Lawrence Paul Rose and Drake Stillman, "The Pseudo-Aristotelian Questions of Mechanis in Renaissance Culture," *Studies in the Renaissance* 18 (1971): 54–104.
24 Pseudo-Aristotle, "Mechanica, III C. BC," in *The Complete Works of Aristotle*, ed. Jonathan Barnes, V. 1 (New Jersey: Princeton University Press, 1984), 1316.
25 Jonathan Sawday, *The Body Emblazoned: Dissection and the Human Body in Renaissance Culture* (London: Routledge, 1995), 111–113, 194–196.
26 Marco Frascari, "The Drafting Knife and the Pen," in *Implementing Architecture: Exposing the Paradigm Surrounding the Implements and the Implementation of Architecture*, ed. Rob Miller (Atlanta: Nexus Press, 1988).
27 Galileo Galilei, *Discoveries and Opinions of Galileo*, trans. Stillman Drake (Norwell: Anchor Press, 1957), 173–216.
28 Galilei, *Dialogues*, 131.
29 Alina Payne, *The Telescope and the Compass: Teofilo Gallacini and the Dialogue between Architecture and Science in the Age of Galileo* (Florence: Olschki, 2012), 75.
30 Galilei, *Dialogues*, 1.
31 Payne, *The Telescope and the Compass*, 55–79.
32 Hart and Hicks, *Paper Palaces*, 173.
33 Alberto Pérez-Gómez and Louise Pelletier, *Architectural Representation and the Perspective Hinge* (New Haven: MIT Press, 2000), 97–125.
34 Massimo Scolari, *Oblique Drawing: A History of Anti-Perspective* (New Haven: MIT Press, 2012), 6–9.
35 Galileo criticized anamorphosis for its distortion of reality (Pérez-Gómez and Pelletier, *Architectural Representation*, 145).
36 Giovanni Antonio Rusconi, *Dell'Architettura* (Venice: Giolitti, 1590), 64.
37 Sebastaino Serlio, *Tutte l'Opere d'Architettura et Prospettiva* (Vicenza: De Franceschi, 1537), Book III, frontispiece.
38 Galilei, *Dialogues*, 130–131. At the anatomical theater in Bologna (1637), a female figure, symbolizing anatomy, and placed above the teacher's desk, is gifted with a thigh bone by an angel. Sawday, *The Body Emblazoned*, 184.
39 James Playfair McMurrich, *Leonardo da Vinci: The Anatomist* (Baltimore: The William and Wilkins Company, 1930), 89–90. Leonard AnA, 18; cf. AnA, 4v; QIV 6.
40 Payne, *The Telescope and the Compass*, 75.
41 Clifford Truesdall, *Essays in the History of Mechanics* (New York: Springer-Verlag, 1968), 22.
42 Giovanni Alfonso Borelli, *De Motu Animalium* (Rome: Angeli Bernabo, 1680), Table 1.
43 Galilei, *Dialogues*, 7.
44 Benvenuto, *An Introduction to the History of Structural Mechanics*, 197, 268–287.

45 Payne, *The Telescope and the Compass*, 62, 68, 142.
46 René Descartes, *Philosophical Essays and Correspondence*, ed. Roger Ariew (Cambridge, MA: Hackett Publishing, 2000 [1641]), 44. John Lewis, "Mersenne as Translator and Interpreter of the Works of Galileo," *Modern Language Notes* 127, 4 (2012): 754–782.
47 Descartes, *Philosophical Essays*, 136–137, 199–200.
48 Descartes, *Philosophical Essays*, 204.
49 Descartes, *Philosophical Essays*, 204.
50 Descartes, *Philosophical Essays*, 178.
51 Jacobus Typotius, *Symbola Divina et Humana* (Prague: Sadeler, 1601–03), 46.
52 Galileo was influenced by William Gilberts' *De Magnete* (1600).
53 B. Castelli,*Discorso sopra la calamita*. Ms Galil. 111, c.203v. Biblioteca Nazionale Centrale, Florence.
54 Vitruvius, *The Ten Books*, X, 284. Vitruvius, *De Architettura*, 1303, 3.
55 http://catalogue.museogalileo.it/object/ModelIllustratingHumanArmAsThirdorderLever.html (accessed on August 16 2015). Bruce Martin, David Burr, Neil Sharkey and David Fyhrie, *Skeletal Tissue Mechanics* (New York: Springer, 1998), 226–227.
56 Joseph Rykwert, "Body and Mind," in *Storia delle Idee Problemi e Prospettive*, ed. Massimo Bianchi (Rome: Edizioni di Ateneo, 1989), 157–168.
57 Julien de la Mettrie, *L'Homme Machine* (Leyden: Elie Luzac, 1748) theorized the body as machine.
58 Mark Johnson, *The Body in the Mind* (Chicago: University of Chicago Press, 1987), XIV–XV.
59 Castelli, *Discorso sopra la calamita*, Ms Galil. 111, c.203v. Biblioteca Nazionale Centrale, Florence.
60 Santiago Calatrava and Cecilia Lewis Kausel, *Santiago Calatrava: Conversations with Students: The MIT Lectures* (Princeton: Princeton University Press, 2002), 43, 47–55, 91–97.

Bibliography

Alberti, Leon Battista. *De re aedificatoria*. Florence: Nicolaus Laurentii, 1485.
Alberti, Leon Battista. *On the Art of Building in Ten Books*. Edited by Joseph Rykwert and Neil Leach. Cambridge, MA: MIT Press, 1991.
Barbaro, Daniele. *Vitruvio. I Dieci Libri dell'Architettura*. Milano: Il Polifilo, 1987 (1567).
Benvenuto, Edoardo. *An Introduction to the History of Structural Mechanics*. New York: Springer-Verlag, 1991.
Borelli, Giovanni Alfonso. *De Motu Animalium*. Rome: Angeli Bernabo, 1680.
Bos, Henk. *Redefining Geometrical Exactness: Descartes' Transformation of the Early Modern Concept of Construction*. New York: Springer, 2001.
Calatrava, Santiago and Cecilia Lewis Kausel. *Santiago Calatrava: Conversations with Students: The MIT Lectures*. Princeton, NJ: Princeton University Press, 2002.
Cesariano, Ceasare. *De Architectura*. London: Benjamin Blom, 1968 (1521).
de la Mettrie, Julien. O. *L'Homme Machine*. Leyden: Elie Luzac, 1748.
Descartes, René. *Philosophical Essays and Correspondence*. Edited by Roger Ariew. Cambridge, MA: Hackett Publishing, 2000 (1641).
Di Giorgio Martini, Francesco. *Trattato di Architettura, Ingegneria e Arte Militare*. Milano: Il Polifilo, 1967 (1474–1482).
Di Pasquale, Salvatore. *L'Arte del Costruire. Tra Conoscenza e Scienza*. Venezia: Marsilio, 2003.
Drake, Stillman. *Mechanics in Sixteenth Century Italy*. Madison: The University of Wisconsin Press, 1969.
Emmons, Paul. "Size Matters: Virtual Scale and Bodily Imagination in Architectural Drawing." *Architecture Research Quarterly*, 9, 2–3 (2005): 227–235.
Emmons, Paul. "A Window to the Soul: Depth in Early Modern Section Drawing." In *Architecture Post Mortem: The Diastolic Architecture of Decline, Dystopia, and Death*, edited by Donald Kunze, David Bertolini and Simone Brott, 153–177. Surrey, UK: Ashgate, 2013.

Favaro, Antonio. *Le Opere di Galielo Galilei*. Firenze: Barbara Editore, 1968.

Filarete, Antonio di Pietro Averlino. *Trattato di Architettura*. Milano: Il Polifilo, 1972.

Frascari, Marco. "The Drafting Knife and the Pen." In *Implementing Architecture: Exposing the Paradigm Surrounding the Implements and the Implementation of Architecture*, edited by Rob Miller, 1–20. Atlanta: Nexus Press, 1988.

Galilei, Galileo. *Dialogues Concerning Two New Sciences*. Translated by Alfonso de Salvio and Henry Crew. New York: Dover Publications, 1954 (1638).

Galilei, Galileo. *Discoveries and Opinions of Galileo*. Translated by Stillman Drake. Norwell: Anchor Press, 1957.

Ginzburg, Carlo. "Microhistory: Two or Three Things That I Know about It." *Critical Inquiry* 20, 1 (1993): 10–35.

Goffi, Federica. *Time Matter(s): Invention and Re-Imagination in Built Conservation: The Unifinished Drawing and Building of St. Peter's in the Vatican*. Surrey, UK: Ashgate, 2013.

Hart, Vaughan and Peter Hicks, eds. *Paper Palaces: The Rise of the Renaissance Architectural Treatise*. New Haven: Yale University Press, 1998.

Johnson, Mark. *The Body in the Mind*. Chicago: University of Chicago Press, 1987.

Latour, Bruno. "How to be Iconophilic in Art, Science and Religion?" In *Picturing Science, Producing Art*, edited by Caroline A. Jones and Peter Galison, 418–439. London: Routledge, 1998.

Leonardo (da Vinci). *Leonardo Da Vinci: Anatomical Drawings from the Royal Library, Windsor Castle*. Edited by Kenneth D. Keele and Jane Roberts. New York: The Metropolitan Museum of Art, 1983.

Lewis, John. "Mersenne as Translator and Interpreter of the Works of Galileo." *Modern Language Notes* 127, 4 (2012): 754–782.

Mallgrave, Harry Francis. *The Architect's Brain: Neuroscience, Creativity, and Architecture*. West Sussex: Wiley-Blackwell, 2011.

McMurrich, James P. *Leonardo da Vinci: The Anatomist*. Baltimore: The William and Wilkins Company, 1930.

Lawrence Rose, Paul and Stillman, Drake. "The Pseudo-Aristotelian Questions of Mechanics in Renaissance Culture." *Studies in the Renaissance* 18 (1971): 54–104.

Palladio, Andrea. *I Quattro Libri dell'Architettura*. Venice: Domenico de' Franceschi, 1570.

Palladio, Andrea. *The Four Books of Architecture*. New York: Dover Publications, 1965 (1570).

Payne, Alina A. *The Telescope and the Compass: Teofilo Gallacini and the Dialogue between Architecture and Science in the Age of Galileo*. Florence: Olschki, 2012.

Pérez-Gómez, Alberto. *Architecture and the Crisis of Modern Science*. Cambridge, MA: MIT Press, 1983.

Pérez-Gómez, A. and Louise Pelletier. *Architectural Representation and the Perspective Hinge*. Cambridge, MA: MIT Press, 2000.

Pseudo-Aristotle. *The Complete Works of Aristotle*. Edited by Jonathan Barnes. New Jersey: Princeton University Press, 1984.

Ramachandran, Vilayanur and Sandra Blakeslee. *Phantoms in the Brain: Probing the Mysteries of the Human Mind*. New York: William Morrow & Company, 1998.

Rose, Lawrence Paul and Stillman Drake. "The Pseudo-Aristotelian Questions of Mechanics in Renaissance Culture." *Studies in the Renaissance* 18 (1971): 54–104.

Rusconi, Giovanni A. *Dell'Architettura*. Venice: Giolitti, 1590.

Rykwert, Joseph. "Body and Mind." In *Storia delle Idee Problemi e Prospettive*, edited by Massimo Bianchi, 157–168. Rome: Edizioni di Ateneo, 1989.

Sawday, Jonathan. *The Body Emblazoned: Dissection and the Human Body in Renaissance Culture*. London: Routledge, 1995.

Scolari, Massimo. *Oblique Drawing: A History of Anti-Perspective*. New Haven: MIT Press, 2012.

Serlio, Sebastiano. *Tutte l'Opere d'Architettura et Prospettiva*. Vicenza: De Franceschi, 1537.

Stevin, Simon. *Les Oeuvres Mathematiques*. Leyden: Elsevier, 1634.

Truesdall, Clifford. *Essays in the History of Mechanics*. New York: Springer-Verlag, 1968.

Typotius, Jacobus. *Symbola Divina et Humana*. Prague: Sadeler, 1601–03.

Vesalius, Andreas. *The Epitome of Andreas Vesalius*. Translated by Levi Robert Lind. Cambridge, MA: The MIT Press, 1969.

Vitruvius. *The Ten Books on Architecture*. New York: Dover Publications, 1960.

Vitruvius. *De Architettura*. Milano: Einaudi, 1997.

Part II
Photography

4 The transformative interface

Fragmentation, process and construction in the photographic representation of architecture

Nigel Green

This chapter looks at the transformative nature of photography and its representation of architecture across a series of technological sites and conceptual frameworks. Photography is described as discipline and techniques that not only records the outcome of the architectural process but is present at all the stages of development from pre-visualisation to afterlife. It represents the physical actuality of the architectural object as a cultural sign – thus facilitating the photographic as part of the process of creation and reflection upon architectural possibilities – and is not simply a means to provide evidence of completion or outcome. Moving on from the idea of 'reproducing reality', understood to be the foundation of photographic representation, this chapter suggests that photographic technologies have evolved to the point at which they are capable of being instrumental in constructing new visual realities. These new realities function to enhance, or challenge, the notion of photography's relation to the 'real' and the conceptualisation of architecture itself. The arguments presented here, then, are premised on photography as a transformative interface that allows ideas to flow bi-directionally. This proposition originates in the key modernist discourses surrounding photography in the 1920s that opposed photographic realism with experimental creativity and invention. It is, however, a proposition also considered in relation to digital technologies of photographic origination and post-production. To place this reading of photography in context, however, it is necessary to return to some of its earliest theories.

The modernist context: An evolution of the architectural image

The photograph interacts with architecture at every stage of the design and building process, from conception and modelling to documentation, promotion and archiving. As Lorenzo Rocha states, "Architecture is assumed to be a cultural product that exists at the intersection of its physical presence and all its visual representations."[1] Since the advent of modernism in the 1920s, the photographic image has become the primary medium through which architecture is presented and disseminated, to the extent that many buildings are only known and experienced through photographic representation. This is not new; the historical relationship between photography and architecture has been mutually productive. The current interest from photographers seeking architectural subjects, and architects themselves becoming more sophisticated in their exploitation of photography, constitutes an active and continued dialogue that is surrounded by a growing culture of critical, historical and contextual re-evaluation.

Architects using photography to promote their work contrast with photographers who look at architecture from a documentary perspective to contemplate a range of historical, ideological and visual ideas. The classical architectural photograph is canonical with established rules and conventions, from perspective control, viewpoint and lighting.[2] In a commercial context, this requires images that represent the moment of a building's completion prior to use – the condition that is closest to that intended by the architect. A format consolidated by the architectural journals and press.[3] Both have traditionally employed the same technical language, which was consolidated in its contemporary form in the 1920s and 1930s by the photography of the Neue Sachlichkeit or 'New Objectivity.' New Objectivity is defined by what Hugo Sieker termed its 'absolute realism,'[4] an approach to photographic representation that formed technically in the 1920s and was championed by the photographer Albert Renger-Patzsch.

The technical and aesthetic language of New Objectivity was employed in conventional architectural photography to create 'spectacular' and 'propositional' images that, in the context of modernism, contributed to the broader discourse of social change and utopianism. Although associated with the 1920s, the language of New Objectivity is still exploited in contemporary photography for its 'neutrality' and ability to produce dispassionate and consistent forms of 'documentary' images that are particularly suited to the recording of architecture and landscape subjects. The archival and conceptual practice of the photographers Bernd and Hilla Becher provides a benchmark example.

The publication of *Anonymous Sculpture* in 1970 represents an approach to the photographic documentation of built space and architecture that continued through 'New Topographics' and the 'Dusseldorf School'. The early work of Thomas Ruff and Andreas Gursky are another example and shared these concerns under the direction of the Bechers themselves. In the context of the work of these photographers, photography can be seen as 'recuperative' – documenting the legacy or 'afterlife' of architecture and operating as a genre that is widely explored in contemporary fine art photography such as Yves Marchand's documentation of dereliction in Detroit.[5]

Contemporary architectural photography largely shares the same visual intentions with that of its modernist predecessors, then, yet, it can be argued, the ideological and cultural context is essentially different. As Owen Hatherley comments with respect to the popular architectural websites such as Dezeen and ArchDaily, they "provide little but glossy images of buildings that you will never visit, lovingly formed into photoshopped, freeze-dried glimmers of non-orthogonal perfection, in locations where the sun, of course, is always shining." Hatherley continues by stating that the denial of physical experience, location and place epitomizes "an architectural culture that no longer has an interest in anything but its own image."[6] Considering this argument as contemporary, however, is not strictly correct. This criticism was also articulated in 1931 when Walter Benjamin quoted Brecht as saying, "Less than ever does the mere reflection of reality reveal anything about reality."

Brecht's statement here is preceded by another observation of interest in understanding the context of contemporary architectural photography – his observation that such photography has more to do with the "salability" of its subject "than of any knowledge it might produce."[7] Architectural photography from the 1920s had an unembellished and functional directness sympathetic to the ideological concerns from which it originated – concerns that were oriented towards the future and the remodelling of social order that can be summed up in in Le Corbusier's words from *Towards a New Architecture* – "Architecture or revolution. Revolution can be avoided."[8] By contrast, the broader socio-economic context we operate in today would seem to suggest that much of contemporary architectural photography is permeated with the visual language of 'lifestyle,' which places emphasis on image, ownership and individual status as opposed to citizenship, social cohesion and collectivism.

There are other lessons that we can draw from considering this historical context comes from the way in which photography undoubtedly influenced the conception and design of modernist architecture in several ways. The exaggeration of monocular perspective with privileged 'photographic' viewpoints becoming built into its formal monochrome purity is one example. Another is modern architecture's monochrome tendencies. Le Corbusier's first housing project in the Bordeaux suburb of Pessac incorporated an external colour scheme, which was lost in the black and white photographs taken of it. As if to confirm Benjamin's observation that "the work of art reproduced becomes the work of art designed for reproduction,"[9] Le Corbusier's later Villa Savoye – and other projects of this period – were devoid of external colour. The coherent formation of modernist architecture during the inter-war period can thus be defined as exactly congruent with the new photography that became synonymous with it. Interestingly, in the context of colour, this was a result of the technological difficulties of colour reproduction.

The new Baryta paper that came into production around the end of the first decade of the twentieth century[10] meant that mass production of high-quality black and white photographs in the form of postcards were possible. These images are known as 'real photograph postcards' to distinguish them from other printing techniques that were also in use at the same time. The history of architectural photography as revealed in archives, journals and monographs then was dominated by these black and white images. Taking the *Architectural Review* as an example, it was not until the mid-1980s that colour was used for approximately 50 per cent of the journal's images, and it was not until the mid-1990s that colour became the standard norm.

In an exhibition at the Bauhaus in 2004 centred on modernist architectural postcards collected by Bernd Dicke and published as the book *Modern Greetings*, the dominance of these black and white image was countered by only three examples of colour reproduction. Ironically, although architects often wanted photographs of their buildings in colour, with many apparently keen users of the Autochrome process, the technical problems this engendered, along with the high cost of production ruled out its practical use; the Autochrome plate was also ill suited for reproduction and panchromatically insensitive to the spectrum of colours that architects were keen to use.[11] Its three-colour printing process required three separately filtered negatives to be made from exactly the same camera position and was thus little used, with the photographers of Bruno Taut's Magdeburg estates for the wall paint company Keim being a rare example.[12]

By contrast, the more populist form of the postcard necessitated colour, and spectacular events such as the World Fair's and Exhibitions demanded an equally spectacular image. The significance of the architectural postcard was its ability to disseminate the new architectural practice to a broader public beyond the industry-specific journal. As Rolf Sachsse points out in his essay for *Modern Greetings*, picture postcards should be seen as "multipliers of new architecture."[13] The postcard is also a transformative process, not only in the dissemination of modernist architecture but also as a technological site for the investigation of colour representation.[14] Whether in colour or black-and-white form, the historical relationship between photography and architecture is one in which the role of the photograph goes far beyond mere 'reproductions' of architectural reality. It is also one in which architecture offered photography a great array of popular content.

Indexing and distorting reality:
The frame and the cut

The discourse between photography and architecture initiated by the historic inception of photography has proved mutually expedient, stimulating the development of each medium to advance the scope of its own technological and visual potential. Owing to the long exposures required to overcome the insensitivity of early photographic materials, the first lens-based chemical images produced by the founding pioneers of the medium, Niépce, Daguerre and Fox Talbot, were of architectural subjects. Fox Talbot notably stated that his image of the Oriel Window at Laycock Abbey, c.1835, was "the first that was ever yet known to have drawn its own picture,"[15] a point that Geoffrey Batchen emphasises by writing of Fox Talbot, "His camera obscura looks out at the inside of the metaphorical lens of the camera of his own house."[16] This innate reflexivity serves to confirm the terminological and structural correspondence between photography and architecture in which the optical device of the camera obscura is simply the duplication of a room or interior. The only addition is a light-sensitive surface with which to record the inverted image.

The camera obscura was used by artists during the Renaissance period, although the extent of its application is disputed.[17] It nonetheless serves to highlight the analogy between the basic perpendicular construction of architecture, its miniaturisation in the camera obscura and its shared demarcation of both three-dimensional and two-dimensional space – the 'frame.' Photography inherited this, and other aspects of its primary visual language, from painting and seamlessly visualised the world according to a set of codes and conventions that were already both culturally and technically embedded. The 'frame' in painting provides the precedent for the 'frame' in photography thus ensuring a contiguous form of representation – a dominant visual referent that has its origins in architecture itself. However, whereas the frame in painting defines a space for representation to be layered or mapped over time, the photographic frame constitutes a 'cut' into an existing field of vision whereby composition neatly excludes peripheral information. It is the mechanical recording or 'indexicality' of the image facilitated by this 'cut' that underpinned photography's privileged claim to the representation of physical reality.

However, the image formed by the camera lens or pinhole is not rectilinear but circular. The image circle in lens-based photography determines the limit of possible movement, the placement of the 'frame,' which allows for perspective correction in conventional architectural photography. Further emphasis is provided by the pinhole, or aperture without a lens, which allows unfocussed light rays to enter the camera obscura at acute angles, producing an image that covers the interior surfaces. The true 'unframed' or 'uncut' image would, therefore, take the form of a circle, thus constituting a radically different form of visual experience and discourse. As Peter Blundell Jones states, "The enthusiasm for perspective correction is surely largely due to the frame, which limits the angle of view, imposes a geometrical composition, and presents both a horizontal and vertical datum with which to compare the content,"[18] thus confirming the syntactical and structural correspondence between architecture and its photographic representation. The language of the perpendicular, the 'frame,' thus suggests that architecture along with painting are complicit in co-opting photography through the imposition of a prior and dominant language.

Although the frame is structurally embedded in photography, the indexical nature of the image – as evidenced by the 'cut' – does, in certain modes of photography such as the 'snapshot,' serve to undermine its visual authority. The canonical architectural photograph, like academic painting, defines itself in relation to the frame as a form of natural construct – a world that is visually coherent and complete and which functions independently of what is unseen beyond the frame. The documentary or snapshot photograph, however, draws attention to the fact that the edges of the frame do not contain, but separate, photographic space from actual space. John Szarkowski expressed this in discussing the specific qualities of the medium in his five 'concepts' of the thing itself, the detail, the frame, time and vantage point. These aspects, he proposed, were most concisely evident in amateur photography and delineated the boundary between painting and photography.[19] In this respect, the photograph must always be a 'fragment,' and its indexing or realism must be seen as something else – if not a distortion, only a partial representation.

Photography and fragmentation – reconstructing histories

Photography evolved as a technological development of modernity that reached a level of maturity as a modernist practice. The fragmentation of experience that accompanied modernity is both reflected in and replicated by photography. Thus photography itself should be viewed as a prime agent of fragmentation. In *The Body in Pieces: The Fragment as Metaphor of Modernity*, Linda Nochlin – referring to Fuseli's chalk and wash drawing, *The Artist Overwhelmed by the Grandeur of Antique Ruins*, 1778–79 – states, "And yet the loss of the whole is more than tragedy. Out of this loss is constructed a distinctively modern view of antiquity as loss – a view, a 'crop,' that will constitute the essence of representational modernism."[20] In her examination of Edouard Manet's, *Masked Ball at the Opera*, 1873, Nochlin identifies the painting's compositional complexity as highlighting the, "significations of the crop, the cut, and the fragmented figure in relation to the representation of modernity and the construction of modernism as a style,"[21] with the painting laying out these conditions "in all their aspects."[22]

In considering the difference between the cutting or cropping of 'pictorial space' and the 'fragmented bodies' created as a result, Nochlin offers three "opposing interpretations."[23] The first being that of 'total contingency,' in which the picture reflects the "meaningless flow of modern reality itself," thus also being devoid of a narrative structure. She relates this to both the realism of nineteenth-century literature and to the 'new medium' of photography with its ability to indiscriminately record visual data. In this reading, the cropped figures at the edge of Manet's painting exhibit the same serendipitous dissection as that created by the photographic frame. In her second interpretation, Nochlin takes the 'total determination' of the aesthetic to be a product of the artist's will. The cut, or crop, becomes a strategy through which the 'device' of the modernist work is made self-evident. Attention is drawn to the "formal organisation of the picture surface" as a "pictorial signifier" and not a "simulacrum of reality."[24]

Finally, Nochlin contemplates the possibility that the cropped borders designate a process of 'image-making as play.' The borders revealing an "oscillation between contingency and determination." Although Nochlin views these as opposing interpretations, it can be argued that they are not exclusive and do in fact combine to express a logic of fragmentation which makes photography central to the process of change and revaluation brought about by the impact of modernity. As Peter Osborne observes in his essay *An Historical Index of Images: The Aesthetic Signification of the Photograph*, "The continuous – or 'all-over' image – imposed by the technical form of the photographic process became a new socio-historically imposed normative form of aesthetic totality to which all other forms – painterly, musical, literary – were tendentially subject."[25]

In looking at the relationship between modernist architecture and its photographic representation in terms of the fragment, and indeed in the terms already laid out in this chapter, it is important to recognise that the relationship is part of an inherited or received space – the knowledge of which is derived from its artefacts. The formation of modernist architecture and space and the discourse that generated it is now a past event. We now encounter its legacy in everyday life as fragments, divorced from the signification of the discourse from which it originated. The photograph functions as a key site through its temporal link to the past as a conveyor of ideology and its physical expression. The photograph configures history at the very point it becomes a severance from it.

History, then, is configured by the agency of chance and is mediated through the subjective act of taking a photograph. The resulting artefact defines a historical space by virtue of its rupture from the continuum of lived time, by becoming a fixed moment, like a layer peeled from the surface that remains in time but also out of time. Thus the photograph is suspended – at an ever-increasing distance – between the contingency of its origin and its consequent afterlife, which, under certain conditions, can form a bridge between the two – a differential other that can re-emerge to connect the past to the present. One example of which is Mies van der Rohe's German Pavilion for the 1929 International Exhibition in Barcelona.

In the chapter *'Mies in Maurelia,'* from Victor Burgin's *The Remembered Film*, Burgin begins with an extract from Italo Calvino's *Invisible Cities* in which the traveller to Maurelia is asked to contemplate the difference between the old provincial city shown in postcard images and the modern metropolis that has replaced it. Concluding that the "old postcards do not depict Maurelia as it was, but a different city which, by chance, was called Maurelia, like this one."[26] Suggesting that the German Pavilion in Barcelona should be viewed in a similar way, Burgin highlights the fact that these observations are determined by the photographic image. "The German Pavilion disappeared in 1930 to reappear in 1986. Photographs are the medium by which it travelled through time – a medium thick with images from the intervening years."[27] As the antique fragment constitutes the loss of a past once whole, it can also be indicative of a future that can be rebuilt. The photographs of the German Pavilion, through which the building in the years of its absence was primarily known and discussed, acted as the agent for its reconstruction.

New vision and the legacy of process

If New Objectivity emphasised the representational qualities unique to the medium of photography with its ability to convey a heightened visual awareness and attention to detail, then New Vision encompasses the avant-garde and experimental photographic practices of the 1920s – from the use of extreme viewpoints, 'above and below,' advocated by Rodchenko to the exploration of 'camera-less' photography pioneered by Moholy-Nagy. Moholy-Nagy was a key exponent of the integration of photography with other forms of art practice – a stance that he developed through his teaching at the Bauhaus. Indeed Moholy-Nagy's position regarding art and design practice as outlined in his book *Painting, Photography, Film, The New Vision* and his last work, *Vision in Motion*, should be viewed as utopian texts comparable in the field of photography and the arts to that of Le Corbusier's *Towards a New Architecture*.

In his short text, "A New Instrument of Vision," Moholy-Nagy writes,

> The photogram, or camera-less record of forms produced by light – which embodies the unique nature of the photographic process – is the real key to photography. . . . The photogram opens up perspectives of a hitherto wholly unknown morphosis governed by optical laws peculiar to itself. It is the most completely dematerialised medium which the new vision commands.[28]

For Moholy-Nagy, the processes of 'analysis-abstraction-extension'[29] lead not only to a rethinking of the medium of photography but also painting, film and architecture. It is the crossover from photography in the form of the photogram that represents the encoding of an anticipatory future realisation in another expanded form that confirms the constructivist photogram as a utopian practice. As Eleanor Hight writes,

> Moholy also used the photogram technique to create suspended objects related to architectural constructions seen in two- and three-dimensional work of others such as El Lissitzky and Gustav Klutsis. Resembling a kind of futuristic space station, the pictorial forms represent the constructivists' proverbial "blueprints" for the architecture of the future.[30]

Thus photography is seen to be not only representative of architecture and architectural space but also as instrumental in providing new visual models for its production.

The photogram also denies single-point perspective, thus establishing a multi-perspectival reading. Related to this is the contrast between 'perspectival' and 'aperspectival' architecture seen in photographic rendering and discussed by Blundell Jones. In this context, Jones compares the 'compelling one-point perspective'[31] of Mies' Barcelona Pavilion with the 'denial of perspective'[32] in the buildings of Hans Scharoun. He states, "For Scharoun, escape from the dominant central axis and symmetry was an escape from both aristocratic and fascist power, leading to a democratic situation where people were permitted equal, but diverse, points of view, both literally and metaphorically."[33] Further to this, with reference to Scharoun's 'Juliet' block of flats in Stuttgart, he argues that the resulting photographs reveal a 'distortion of perspective,' which problematizes its reading as a photographic image.

Both aperspectival architecture and the photograms of Moholy-Nagy, then, serve to highlight the limitations of monocular, single-point perspective to represent other forms of spatial organisation. Through this, the complicity between the photographic apparatus and its inherited conventions of representation are revealed. The absence of a referent in the camera-less, non-representational form of the photogram constitutes a radical break with convention, allowing photography to engage in the creation of new spatial order as well as simply recording it in actuality. From this perspective of an expanded definition of photographic practice, Moholy-Nagy could write,

> Through photography, we can participate in new experiences of space. . . . With their help, and that of the new school of architects, we have an enlargement and sublimation of our appreciation of space, the comprehension of a new spatial culture.[34]

Moholy-Nagy was not alone in developing the photogram as an autonomous modernist practice. Christian Schad and Man Ray also developed their own variations of the process. The photogram is not, however, unprecedented, with the process, in the form of contact printing, accounting for some of the earliest examples of photography. Fox Talbot's first experiments with light-sensitive surfaces simply reveal the imprint of objects placed upon them such as lace or plant material. Gail Buckland, in her book *Fox Talbot and The Invention of Photography*,[35] reproduces, unfortunately without colour, an early example of Fox Talbot's work, which bears no visible image. Buckland states that many of Fox Talbot's early experiments exist in similar states. The experimental nature of this 'photograph' suggests other potential readings of the 'process.' We are presented with the evidence of a chemical reaction but no image is manifested. The reading of this and other similar pieces should perhaps be more closely linked to the 'desire'[36] to capture an image of the world. In a sense, Fox Talbot's photogenic drawing with no discernible image functions as a figure of the imaginary, which encodes the desire for representation itself. The desire survives in the chemical stain of error, whereas the object of representation does not.

Underpinning Moholy-Nagy's conception of photography was the notion that the mistake or error was an essential part of the medium that could be utilised and developed. This can be seen in relation to the polarisation of the paradigm of photography as constitutive of a process producing specific, technically accomplished results and the alternative morphological and contingent aspects signified by imperfection and chance. Thus the photogram as the construction of controlled chance were seen as producing images that are both virtual and propositional. The 'desire' inherent in the limitations and failure of Fox Talbot's early experiments to achieve a technically perfected form of representation can be said to be mirrored in the constructivist photogram's predictive 'desire' to project the possibility of a new 'spatial culture' and social order.

The constructed image in the digital age

The desire to modify or change the photograph in order to communicate a specific meaning or idea is intrinsic to the photographic representation of architecture. The technical and aesthetic limitations of photography usually coalesce around its ability to record too much or too little information. With digital imaging technology, the transformation of the photograph takes on new, seamless and less transparent forms. Le Corbusier is well-known for his doctoring of photographs to make them conform to his ideological intent. Linking visual representation to conceptual integrity, he saw photography as being no more than a means to an end, in which the "designers of High Modernism – should not only compose the material world but, first of all, constantly organize thinking itself."[37] In her book *Privacy and Publicity*, Beatriz Colomina writes that of the photographs of Le Corbusier's early Villa Schwob included in the journal, *L'Esprit Nouveau 6*, he "air-brushed the photographs. . . . to adapt them to a more 'purist' 'aesthetic.'"[38] She then identifies the different parts within the photograph that were modified or erased, such as the rationalisation of structural elements and the removal of "organic growth" and "distracting objects."

The active cleansing of the photograph to reveal or construct an alternative and uncompromised visual space represents the exclusion of the documentary principle that underpins photography. This equally articulates the modus operandi of conventional or commercial architectural photography in which arbitrary and random traces of human presence are to be expurgated. To further this point, Beatriz Colomina makes reference to the architectural theorist Stanislaus von Moos, where he states that for Le Corbusier, "Architecture is a conceptual matter to be resolved in the realm of ideas, that when architecture is built it gets mixed with the world of phenomena and necessarily loses its purity."[39] Colomina continues by suggesting that built architecture is only part of a process, and only when it "enters the two-dimensional space of the printed page" does it return "to the realm of ideas."[40] For Le Corbusier, the essential role of photography was to provide a conceptual space in which ideas could be revised and developed, thus presenting an ideal visual narrative that runs in parallel to the phenomenal world.

The photographic representation of architecture is constructed in different ways, from the conventions of framing, process manipulation and conceptual reinvention. The consistent factor is that the architectural idea is evidenced visually. Whether the technologies that underpin the process of representation are self-evident or not, we may agree that the visual signifier conveys the idea of architecture through the forms we associate with it. With the advent of digital photography and post-production, the boundary between notions of architectural documentation, and other forms of photo-based architectural representation, becomes open to further questioning and interpretation.

Andreas Gursky's photograph, *Paris, Montparnasse*, 1993, shows the monumental façade of the Mouchotte building in Paris in a comparatively monumental printed format of over four meters in width. Seemingly representative of the conventions embodied by the post-Becher, Düsseldorf School of New Objectivity – with its use of large format photography, corrected perspective and flat, overcast lighting – there is also something 'unphotographic' that destabilises this reading.

The ability to represent extreme levels of detail and resolution at this scale contradicts the established language of photographic formats. In order to achieve this outcome, Gursky has joined two photographs, thus producing one of his first digitally manipulated images, which also established a precedent for his ongoing practice. The simulation of objective documentation through digital manipulation establishes a paradigm shift within the language and conventions of New Objectivity. The visual sign articulates the discourse of New Objectivity and documentary photography, yet it is a construct or fictional composite. Gursky has continued to exploit the potential of using multiple photographs to construct new and spectacular visual interpretations of architecture. The result is a suspension of the referent in a kind of digitised limbo, somewhere between the real and the imaginary. The technical advances in the digital medium make the production of such photographs commonplace across the multiple software applications that are used daily by amateur and professional photographers alike. Such post-production techniques have become so embedded in the language of photography that their integrity and use are rarely questioned or challenged.

The desire to create images from multiple photographs is not new in photography, as the early use of combination printing techniques by the photographers O. Rejlander and Henry Peach Robinson demonstrate. Their intention was also to use photographic representation to both construct artistic vignettes and overcome the technical limitations of the medium. The production of separate images for different parts of the composition would be printed separately onto a piece of paper by exposure to daylight. Significant skill was required to produce a seamless image. The result was an immediate challenge to the notion of photographic truth. Photography's relationship to the representation of reality by mechanically recording an image without artistic interpretation forms the ontological foundations of the medium. Claims for photography to be considered as an art form thus constituted a critical battleground that has played out through most of photography's history and development, and it can be argued that photography has only recently succeeded in achieving this status with photographic artists such as Gursky.

If the aforementioned examples of constructed photography maintain an ambiguity in the relationship between reality and photographic representation– i.e. its status as art – the work of Filip Dujardin emphatically exploits the fictional potential of digital imaging to question the nature of both photography and architecture. His series, *Fictions* are composite digital collages that play with tropes of late modernist architecture to create images of new, un-built and un-buildable structures. Unlike the constructivist photogram, Dujardin's work is not propositional, but rather constitutes a critical frame that interrogates the legacy of the built environment. Pedro Leão Neto states that Dujardin's work is "characterized by an artistic strategy that uses the tools of digital imaging as an opportunity for a postmodern project that unveils the pretence of photographic objectivity,"[41] thus making explicit what remains implicit in the use of photography by Le Corbusier and in the early digital combination images of Andreas Gursky – a questioning of photographic objectivity through the adoption of the aesthetic formalism of New Objectivity which reveals the internal contradictions of a language centered on the communication of the medium's visual purity.

Dujardin's *Fictions* constitute a form of hybrid photography where multiple image fragments are reconfigured into a single composite. Photographic fragmentation is duplicated as an internal visual mechanism, which stages a confrontation between the perceptual and conceptual nature of the image.

His use of digital manipulation, Neto suggests, does not "mask or denature a profound reality," but calls "our critical attention to it, and his work is used, not to blur our ontological distinctions between the imaginary and the real, but to sharpen our critical attitude towards the existing architecture of today"[42] (Fig. 4.1).

Conclusion

In addressing the question of how if photographic technologies are instrumental in constructing new visual realities that function to enhance or challenge the notion of photography's relation to the 'real' on the one hand and examining the conceptualisation of architecture itself on the other, we can see that the notion of the 'real' is a construct of the visual codes inherent in the apparatus and language of photography itself. It is a reading exemplified by monocular perspective, the convention of the frame and resulting visual and temporal fragmentation. It is also a reading that suggests intent determines outcome by exploiting the naturalism or indexical nature of photography as proof, or demonstration, that the conceptual content corresponds to actual content.

Figure 4.1 Untitled from series 'Fictions', 2007, Filip Dujardin

In looking at the relationship between modernist architectural space and its photographic representation to explore this, we have seen that the two are intrinsically bound to the progress of modernity prior to the dissolution of social ideologies by the homogeneity of capitalism. The encoding of time and cultural memory in the technology of photographic processes changed how we think about the representation and conception of architectural history and taught us that we should not only think about what is represented but also how that representation is made manifest.

These thoughts are evident in our closing commentary on the recent practices which engage with the legacy of modernist architecture to highlight different concerns. Here photography can be seen as a transformational process whereby ideas are explored and communicated. Digital technology has enabled contemporary lens-based investigation of architecture to be extended further into the realms of invention and visual reinterpretation – albeit in ways that have historical precedents in the form of 'outmoded' techniques and processes, as we have seen. In this respect, the photographic image is a fiction conditioned by the technological possibilities and the cultural conventions of the present. Photography, then, must be seen as never a stable or fixed entity, but rather a process that is continually subject to technological innovation and change. The consistent aspect of photographic representation is that it is always a conduit through which ideas can take on visual form and, as such, it should continue to embrace its potential as a critical tool that both reflects upon and interrogates the process and legacy of architectural discourse.

Notes

1 Lorenzo Rocha, "Building Architectural Images: On Photography and Modern Architecture," in *Concrete: The Photography of Architecture*, ed. Daniela Janser, Thomas Seelig, and Urs Stahel (Zurich: Scheidegger and Spies, 2013), 47.
2 James S. Ackerman argues that the historical development of architectural photography and its visual conventions are based on the replication of architectural drawing. James S. Ackerman, "On the Origins of Architectural Photography," 2 *CCA Mellon Lectures* (2001). Accessed December 19, 2015. http://www.cca.qc.ca/system/items/1937/original/Mellon02-JA.pdf?1241159343
3 The post-war commercial conventions of architectural photography, Nicholas Olsberg argues, represent a rejection of the radical viewpoints and experimentalism of the 1920s in favour of a return to "the emptiest and most static conventions of picturing architecture. This led to decades of the commercial photographer's neatly framed views of uninhabited buildings, set under a blue sky, devoid of passing traffic, lit inside by relentless floods of artificial light, and peopled, if at all, like Julius Shulman's models in Koenig's Case Study 22: in vividly complementary dress and posture". Nicholas Olsberg, "Shattered Glass: The History of Architectural Photography," *The Architectural Review* (2013). Accessed December 19, 2015. http://www.architectural-review.com/rethink/viewpoints/shattered-glass-the-history-of-architectural-photography/8656969.fullarticle
4 Hugo Sieker, "Absolute Realism: On the Photographs of Albert Renger-Patzsch," in *Photography in the Modern Era: European Documents and Critical Writing 1913–1940*, ed. Christopher Phillips (New York: Aperture, 1989), 110.
5 Yves Marchand and Romain Meffre, *The Ruins of Detroit* (Gottingen: Steidl, 2010).
6 Owen Hatherley, "Photography and Modern Architecture," *Photographers' Gallery Blog* (2012). Accessed April 20, 2015. http://thephotographersgalleryblog.org.uk/2012/12/10/photoarchitecture1/
7 Walter Benjamin, "Little History of Photography," in *Selected Writings Volume 2 (1927–1934)*, ed. Michael W. Jennings, Howard Eiland, and Gary Smith (Cambridge: The Belknap Press of Harvard UP, 1999), 526.
8 Le Corbusier, *Towards a New Architecture* (London: The Architectural Press, 1948), 268–269.
9 Walter Benjamin, "The Work of Art in the Age of Mechanical Reproduction," in *Illuminations*, ed. Hannah Arendt (New York: Schocken Books, 1969), 226.
10 Baryta is a barium sulphate coating that facilitates greater image definition and tonal range.
11 Kristen Baumann, *Modern Greetings: Photographed Architecture on Picture Postcards 1919–1939*, comp. Bernd Dicke, and ed. Rolf Sachsse (Stuttgart: Arnoldsche Verlagsanstalt, 2004), 190.
12 Baumann, *Modern Greetings*, 190.
13 Baumann, *Modern Greetings*, 182.
14 The black-and-white postcard is a photographic and not a reprographic reproduction. Prior to economically viable reproductive technologies, the representation of colour was a key problem for photography. Consequently, a variety of solutions were employed from the hand colouring of black-and-white postcards through to hybrid type images that combined monochromatic dot screen prints with added layers of printed colour.
15 Geoffrey Batchen, "Desiring Production," in *Each Wild Idea: Writing, Photography, History* (Cambridge, MA: MIT Press, 2001), 9.

16 Batchen, "Desiring Production," 9.
17 For further reference see David Hockney and Charles M. Falco, "Optical Insights into Renaissance Art," *Optics and Photonics News* (July 2000). Accessed April 14, 2015. http://fp.optics.arizona.edu/SSD/art-optics/papers/OPN.pdf
18 Peter Blundell Jones, "The Photo-Dependent, the Photogenic and the Unphotographable: How Our Understanding of the Modern Movement Has Been Conditioned by Photography," in *Camera Constructs: Photography, Architecture and the Modern City*, ed. Andrew Higgot and Timothy Wray (Farnham, England: Ashgate, 2012), 49.
19 John Szarkowski, "The Photographer's Eye," *Introduction to the Catalogue of the Exhibition* (1964). Accessed May 25, 2015. http://www.jnevins.com/szarkowskireading.htm
20 Linda Nochlin, *The Body in Pieces: The Fragment as a Metaphor of Modernity* (New York: Thames and Hudson, 1994), 8.
21 Nochlin, *The Body in Pieces*, 36.
22 Nochlin, *The Body in Pieces*, 36.
23 Nochlin, *The Body in Pieces*, 37.
24 Nochlin, *The Body in Pieces*, 37.
25 Peter Osborne, "An Historical Index of Images: The Aesthetic Signification of the Photograph." Essay for the exhibition *Ruins in Reverse: Time and Progress in Contemporary Art*, at CEPA Gallery, Buffalo, NY (1998–99). Accessed June 2007. www.cepagallery.com
26 Victor Burgin, "Mies in Maurelia," in *The Remembered Film* (London: Reaktion Books, 2004), 75.
27 Burgin, "Mies in Maurelia," 86.
28 Laszlo Moholy-Nagy, "A New Instrument of Vision," in *Moholy-Nagy: Documentary Monographs in Modern Art*, ed. Richard Kostelannetz and Allen Lane (London: The Penguin Press, 1971), 50.
29 Moholy-Nagy, "A New Instrument of Vision," 93.
30 Eleanor M. Hight, *Picturing Modernism: Moholy-Nagy and Photography in Weimar Germany* (Cambridge, MA: MIT Press, 1995), 111.
31 Jones, "The Photo-Dependent," 49.
32 Jones, "The Photo-Dependent," 53.
33 Jones, "The Photo-Dependent," 53.
34 Hight, *Picturing Modernism*, 78.
35 Gail Buckland, *Fox Talbot and the Invention of Photography* (London: Scolar Press, 1980), 28.
36 For an account of the complex issues surrounding the invention of photography, see Geoffrey Batchen, *Burning with Desire: The Conception of Photography* (Cambridge, MA: MIT Press, 1999).
37 Andrzej Piotrowski, "Le Corbusier and the Representational Function of Photography," in *Camera Constructs: Photography, Architecture and the Modern City*, ed. Andrew Higgot and Timothy Wray (Farnham: England: Ashgate, 2012), 39.
38 Beatriz Colomina, *Privacy and Publicity: Modern Art as Mass Media* (Cambridge, MA: MIT Press, 1996), 107–111.
39 Colomina, *Privacy and Publicity*, 114–118.
40 Colomina, *Privacy and Publicity*, 114–118.
41 Pedro Leão Neto, "Fictions by Filip Dujardin," *Scopio*. Accessed May 5, 2015. http://www.scopionetwork.com/node/244#1
42 Leão Neto, "Fictions".

Bibliography

Ackerman, James S. "On the Origins of Architectural Photography." *2 CCA Mellon Lectures* (2001). Accessed December 19, 2015. http://www.cca.qc.ca/system/items/1937/original/Mellon 02-JA.pdf?1241159343

Batchen, Geoffrey. *Burning with Desire: The Conception of Photography*. Cambridge, MA: MIT Press, 1999.

Batchen, Geoffrey. "Desiring Production." In *Each Wild Idea: Writing, Photography, History*, 2–24. Cambridge, MA: MIT Press, 2001.

Baumann, Kristen. *Modern Greetings: Photographed Architecture on Picture Postcards 1919–1939*. Compiled by Bernd Dicke, and edited by Rolf Sachsse. Stuttgart: Arnoldsche Verlagsanstalt, 2004.

Benjamin, Walter. "Little History of Photography." In *Selected Writings Volume 2 (1927–1934)*, edited by Michael W. Jennings, Howard Eiland, and Gary Smith, 507–530. Cambridge, MA: The Belknap Press of Harvard UP, 1999.

Benjamin, Walter. "The Work of Art in the Age of Mechanical Reproduction." In *Illuminations*, edited by Hannah Arendt, 217–252. New York: Schocken Books, 1969.

Buckland, Gail. *Fox Talbot and the Invention of Photography*. London: Scolar Press, 1980.

Burgin, Victor. "Mies in Maurelia." In *The Remembered Film*, 74–88. London: Reaktion Books, 2004.

Colomina, Beatriz. *Privacy and Publicity: Modern Art as Mass Media*. Cambridge, MA: MIT Press, 1996.

Hatherley, Owen. "Photography and Modern Architecture." *Photographers' Gallery Blog* (2012). Accessed April 20, 2015. http://thephotographersgalleryblog.org.uk/2012/12/10/photoarchitecture1/

Hight, Eleanor M. *Picturing Modernism: Moholy-Nagy and Photography in Weimar Germany*. Cambridge, MA: MIT Press, 1995.

Hockney, David and Charles M. Falco. "Optical Insights into Renaissance Art." *Optics and Photonics News* (July 2000). Accessed April 14, 2015. http://fp.optics.arizona.edu/SSD/art-optics/papers/OPN.pdf

Jones, Peter Blundell. "The Photo-Dependent, the Photogenic and the Unphotographable: How Our Understanding of the Modern Movement Has Been Conditioned by Photography." In *Camera Constructs: Photography, Architecture and the Modern City*, edited by Andrew Higgot and Timothy Wray, 47–60. Farnham, England: Ashgate, 2012.

Leão Neto, Pedro. "Fictions by Filip Dujardin." *Scopio*. Accessed May 5, 2015. http://www.scopionetwork.com/node/244#1

Le Corbusier. *Towards a New Architecture*. London: The Architectural Press, 1948.

Marchand, Yves and Romain Meffre. *The Ruins of Detroit*. Gottingen: Steidl, 2010.

Mitchell, William J. *The Reconfigured Eye: Visual Truth in the Post-Photographic Era*. Cambridge, MA: MIT Press, 1992.

Moholy-Nagy, Laszlo. "A New Instrument of Vision." In *Moholy-Nagy: Documentary Monographs in Modern Art*, edited by Richard Kostelannetz and Allen Lane. London: The Penguin Press, 1971.

Nochlin, Linda. *The Body in Pieces: The Fragment as a Metaphor of Modernity*. New York: Thames and Hudson, 1994.

Olsberg, Nicholas. "Shattered Glass: The History of Architectural Photography." *The Architectural Review* (2013). Accessed December 19, 2015. http://www.architectural-review.com/rethink/viewpoints/shattered-glass-the-history-of-architectural-photography/8656969.fullarticle

Osborne, Peter. "An Historical Index of Images: The Aesthetic Signification of the Photograph." Essay for the exhibition *Ruins in Reverse: Time and Progress in Contemporary Art*, at CEPA Gallery, Buffalo NY, 1998–99. Accessed June 2007. www.cepagallery.com

Piotrowski, Andrzej. "Le Corbusier and the Representational Function of Photography." In *Camera Constructs: Photography, Architecture and the Modern City*, edited by Andrew Higgot and Timothy Wray. Farnham, England: Ashgate, 2012.

Rocha, Lorenzo. "Building Architectural Images: On Photography and Modern Architecture." In *Concrete: The Photography of Architecture*, edited by Daniela Janser, Thomas Seelig, and Urs Stahel, 47. Zurich: Scheidegger and Spies, 2013.

Sieker, Hugo. "Absolute Realism: On the Photographs of Albert Renger-Patzsch." In *Photography in the Modern Era: European Documents and Critical Writing 1913–1940*, edited by Christopher Phillips, 110–115. New York: Aperture, 1989.

Szarkowski, John. "The Photographer's Eye." *Introduction to the Catalogue of the Exhibition* (1964). Accessed May 25, 2015. http://www.jnevins.com/szarkowskireading.htm

5 'The great publicist of modern building'

Photography and architecture in the inter-war years

Valeria Carullo

In the November issue of 1932, the *Architectural Review* published two photographs of seemingly contemporary buildings next to each other on a page that otherwise included only short captions at the bottom (Fig. 5.1). What the two images have in common is a strong vertical thrust, but while the subject of the image on the left, a worm's-eye view of the recently completed no. 500 Fifth Avenue in New York by Shreve Lamb and Harmon is immediately recognisable, the structure on the right is not, framed as it is in a nearly abstract composition. The effect of disorientation is calculated, as the author of the caption underlines by writing, "A new form of sky-scraper in the grand manner? No! This seemingly splendid verticality of glass and steel is none other than an unfamiliar bird's eye view of the roof of our old friend Victoria Station, London."[1] The following January, the journal published a letter by the architect Basil Ionides, who not only criticised the editor for the excessive space given to photography – "I would suggest that your paper should be re-christened 'The Photographic Review'" – but also expressed his concern at the effect some of these images might have: "At the moment the danger is that young architects might try to design buildings that look to the human eye as the photographs shown on page 219 of your November issue."[2] Ionides' comment is a very early example of the realization, within the architectural community, that photography was not only a powerful tool in the communication and promotion of architecture, but could actually have an impact on architectural design itself. Since then, this recognition has become a widely accepted theory, and particular emphasis has been put on the influence that the New Photography of the 1920s and 1930s might have had in the development of the International Style and on modern architecture in general. That page in *Architectural Review* of 1932 encapsulates many of the motifs of the New Photography: extreme angles, abstraction, the theme of comparison and contrast, the use of an image 'out of context' – not to mention the explicit reference to the idea of the 'unfamiliar' view, which was central to the 'new vision' promoted by László Moholy-Nagy.

The New Vision, or Neues Sehen, had its roots in the technological culture of the early twentieth century[3] and at its heart Moholy-Nagy's writings and his work at the Bauhaus. Moholy-Nagy took up photography in 1922 and a year later started teaching at the Bauhaus, integrating photography into the school's general curriculum. He championed a dynamic and creative approach to the medium, which included the use of unusual viewpoints – such as bird's- and worm's-eye views – radical cropping and extreme close-ups, all devices that led towards abstraction. Moholy-Nagy was interested in discovering the camera's unique 'way of seeing', independent from that of the other visual arts; as he states in the Bauhaus book of 1925 *Malerei Fotografie Film* (Painting, Photography and Film),

Figure 5.1 500 Fifth Avenue and Victoria station, London. *Architectural Review*, November 1932, p. 219, Architectural Press Archive / RIBA Collections

The camera has offered us amazing possibilities, which we are only just beginning to exploit . . . for – although photography is already over a hundred years old – it is only in recent years that the course of development has allowed us to see beyond the specific instance and recognise the creative consequences.[4]

Photography was no longer seen as a tool of representation but as an opportunity for creation. A number of heterogeneous influences had helped shape this 'new vision': the advances in camera technology, especially the introduction of new optics, which produced images of unprecedented clarity[5] and light-weight cameras such as the Leica, which allowed greater flexibility and freedom in the choice of viewpoint; visual art movements such as Cubism, Expressionism and Constructivism; the widespread use of aerial photography during the war, which had introduced a new way of looking at the city and the built environment in general;[6] and the new status acquired by cinema as the ultimate modern art. Moholy-Nagy also experimented with and encouraged the use of photo-collage, photomontage and abstract photograms.[7] Photography at the Bauhaus was in fact often combined with modern typography and playful graphics in the pages of books, magazines and posters; all these techniques were employed to instigate a fresh, new way to look at the world.

In Russia at the same time, artists such as Alexander Rodchenko and El Lissitzky saw in photography the most appropriate way to represent the changes taking place in their country after the Revolution. According to Rodchenko, "Only the camera seems to be really capable of describing modern life."[8] Like Moholy-Nagy and his colleagues at the Bauhaus, Rodchenko and El Lissitzky intended to break old habits of perception and visual representation by adopting similar devices in their photography – extreme angles, distorted perspectives, tilted horizons, unconventional close-ups – as well as the use of photomontage. Rodchenko believed that

> in order to educate man to a new vision, everyday familiar objects must be shown to him with totally unexpected perspectives and in unexpected situations. New objects should be depicted from different sides in order to provide a complete impression of the object.[9]

A somewhat different interpretation of the role of photography was advanced by the artists of the Neue Sachlichkeit, or New Objectivity. What they had in common with the proponents of the New Vision was an interest in the camera as a new and independent means of expression. They aimed, however, at capturing reality with a direct, dispassionate, clinical outlook which precluded the more experimental techniques adopted by Moholy-Nagy or Rodchenko. Photographers such as August Sander and Karl Blossfeld were interested in the 'essence' of the object and, as a result, their images were rigorously composed and often in very sharp focus. Their analysis of the object also revealed to them the importance of the detail, which could in some cases be more eloquent than the whole at expressing the essence of the object. Very similar aims were pursued by artists such as Paul Strand and Edward Weston in the United States, who were representatives of the aptly named Straight Photography.

Architecture, photography and art converge

In 1929, two major exhibitions acknowledged the importance of the New Photography: *Fotografie der Gegenwart* (Photography of the Present) in Essen and the better known *Film und Foto*, organised by the Deutsche Werkbund in Stuttgart. Both exhibitions subsequently travelled to other venues in Germany or abroad. In *Film und Foto*, the artistic ideas of the New Vision were showcased in the work of the European photographers selected by Moholy-Nagy and Siegfried Giedion, while its principles were enunciated in the same year by the historian and art critic Franz Roh in an essay whose title, 'Foto-Auge' (Photo-Eye),[10] once again stressed photography's independence of vision. New Objectivity and Straight Photography were also well represented in the exhibition, which counted among its many visitors Charles Sheeler, the American painter and photographer. Sheeler, whose work was influenced by Cubism as well as the moving image, had two years earlier produced the well-known series of photographs of the Ford Plant at River Rouge near Detroit, arranged in a cinematic sequence ranging from panoramic views to semi-abstract details. These images reflected avant-garde artists' fascination with the machine age and their belief in the endless possibilities industrialisation seemed to offer, which was suitably expressed by a modern artistic medium which was itself a product of modern technology. Sheeler, whose photographic work focused primarily on architectural subjects, had in 1920 celebrated the modern city by capturing New York skyscrapers both in photographs and film;[11] his still images, taken from an elevated position, are characterised by high tonal contrasts and very tight cropping that virtually excluded the line of the horizon. Later in the decade, architect Eric Mendelsohn, during a trip to the United States, found himself looking at skyscrapers from street level; like others trying to capture the vertiginous height of these structures, he tilted the camera upwards, abandoning the straight verticals which had been until then a prerequisite of 'correct' architectural photography. These dynamic views – which, as Robert Elwall remarks, "well conveyed the breathless excitement the architect first felt on walking through the streets of New York, Chicago and Detroit"[12] – were published as a cinematic sequence in his 'architect's picture book' *Amerika*[13] of 1926, which received the enthusiastic praise of El Lissitzky:

> Leafing through the pages for the first time grips us like a dramatic film. Completely strange pictures unwind before our eyes. You have to hold the book over your head and twist it around to understand some of the photographs. The architect shows us America, not from the distance, but from the inside; he leads us through the canyons of its streets.[14]

Mendelsohn's book played an important role in introducing American steel construction and tall buildings to a European audience; its illustrations were equally influential, echoing with their spontaneous dynamism the experimental approach of the Bauhaus artists.

It could reasonably be claimed that the Bauhaus was the place where the alliance between modern architecture and modern photography was first forged. As the vocabulary of the new architecture was being developed, there was a cross-fertilisation between the areas of design and photography, especially because the Bauhaus artists did not see them as separate fields. As Robert Elwall pointed out, there is a clear correspondence between the New Photography's belief in exploiting the medium's unique qualities and modern architecture's emphasis on the explicit expression of structure and materials. The simple, bold forms and smooth surfaces – steel, concrete, glass – of the new architecture provided the perfect subject material for a type of visual representation that favoured geometrical compositions, contrasts of light and shadow and strong diagonals. Reflective surfaces added further creative possibilities for the photographer, together with neon illumination, which made capturing architecture at night possible and exciting. Not surprisingly, some of the most striking images of modern architecture of the 1920s were taken by German photographers, in particular by two proponents of the New Objectivity, Werner Manz and Albert Renger-Patzsch. They believed that only photography could give them the tools to achieve their main aim – the representation of reality with clarity and precision. Werner Manz's sharply focused images, often composed with few essential elements and not unfrequently veering towards abstraction, are among the most powerful and evocative of inter-war Modernism. Richard Pare highlighted an essential quality in the German photographers' work, which could be extended to other photographers of that era:

> There is an empathy with the architect's intention. The solution is purely photographic and utilizes the particular characteristics of photography to further the spirit of the buildings. Through this close collaboration, a sense of the period is instilled in photographs; they can be looked upon as reflections on architecture, rather than mere pictures of architecture.[15]

Let us keep this concluding observation in mind, as it is crucial to a further analysis of the influence of the photography on architecture.

The close interrelation between architecture and photography in the 1920s and 1930s was not restricted to Germany and the Bauhaus. In other European countries, practitioners from both fields shared an optimistic outlook on the modern world – a new world that they believed they were contributing to creating – and inspired each other's vision. This was especially true in Czechoslovakia, one of the centres of the Modern Movement, as well as a country where artists "embraced photography and cinema as contemporary media that embodied the spirit of modernity".[16] It is here that one of the first opportunities for reflection on the relationship between architecture and photography was produced: the collection of essays Život; Sborník nové krásy (Life; Anthology of the New Beauty)[17] published in 1922 by the avant-garde group Devětsil. This publication, which counted Le Corbusier among its contributors and gave ample space to photographic illustrations, celebrated photography and cinema as "potent agents of modern technology that promised to transform art";[18] its editor, architect Jaromír Krejcar, described photography as "the only interpreter of the new beauty",[19] while artist Karel Teige compared the beauty of a photograph to that of "an airplane, ocean liner and electric bulb"[20] in his seminal essay Foto Kino Film, which prefigures the themes of the later and better known Malerei Fotografie Film. Czechoslovakia also gave birth to one of the most talented and daring photographers of the period, Jaromír Funke, whose images of architecture are imbued with the principles of the New Vision.

Commercial photography and the power
of the reproduced image

How extensive, however, was the adoption of the language of the New Photography in the representation of buildings in the inter-war period? The 1930s saw the explosion of commercial enterprises specialising in this field, and many of them absorbed the innovations introduced by avant-garde photographers, albeit to varying degrees and often in diluted forms. Bird's- and worm's-eye views were largely adopted, but they were no longer the result of the dynamic approach afforded by handheld cameras, or the difficulties of photographing very tall buildings; they were in fact taken with large format cameras on tripods. In addition, the angled view was often conceived by the photographer as one element of a sequence that included more traditional exteriors and interiors, as well as details. This 'adjustment' has often been interpreted as the reduction of a creative, highly expressive device to a sterile, purely formal one. It could instead be argued that, by introducing the angled perspective to the repertoire of the commercial photographer, the New Vision had succeeded in broadening the boundaries of the standard, accepted view of the architectural object, and the effect that this had on the collective visual consciousness cannot be underestimated. What has *now* perhaps become a tired formula – and yet still extremely effective in conveying a sense of height – was, in the inter-war years, part of a newly acquired expressive range that not only enriched the photographer's vocabulary but allowed him to highlight some of the essential characters of the new architecture.

As both Robert Elwall and the authors of *Architecture Transformed*[21] – Cervin Robinson and Joel Herschman – have pointed out, it was in the countries where modern architecture made a late appearance, Britain and the United States, that a more innovative language in commercial architectural photography was developed. It was probably the need to 'sell' the Modern Movement to a still unconvinced audience that helped in shaping a very distinctive and seductive style,[22] exemplified in the States by the images produced by F. S. Lincoln and Hedrich-Blessing, and in Britain by the work of Dell & Wainwright, official photographers of the *Architectural Review* between 1930 and 1946. In the early 1930s, Dell & Wainwright photographed extensively Art Deco as well as 'traditional' architecture, but it is their dramatic images of modern buildings bathed in sunlight, on the background of cloud-flecked skies, made darker by the use of red and orange filters to accentuate the contrast with the often seemingly white walls, that were etched in readers' memory and are still considered their trademark creation (Fig. 5.2). As Robert Elwall puts it, "Their images leap from the page with an almost irresistible appeal and power."[23] Their highly influential photographic language is, however, the result not only of their talent and peculiar sensibility but also of the close collaboration with the editors of the *Review*. The journal was a very vocal promoter of Modernism and employed Dell & Wainwright's imagery as part of a campaign aimed at popularising the new architecture not only within the profession but also with a wider public. To this purpose, the editor Hubert de Cronin Hastings and his collaborators reshaped the journal by adopting new, creative graphics and page layouts, extensive use of photographic images often in large format and articles by eloquent supporters of the Modernist cause. Hastings' precept that general views should be kept small but details large reflected the newly found significance of the fragment within photographic representation and the belief that a well-chosen detail

Figure 5.2 Miramonte, Kingston-upon-Thames, London (1937). Architect: E. Maxwell Fry. Photographers: Dell & Wainwright. Dell & Wainwright / RIBA Collections

(Fig. 5.3) could stand for the whole building, or even reveal more of it than a general view. It was on the pages of the *Review* that critic Philip Morton Shand commented on the role played by the New Photography in the affirmation of the new architecture: "Did modern photography beget modern architecture or the converse? It is an interesting point. But since their logical development was simultaneous, and their interaction considerable, it hardly matters which."[24] Shand also stated that "without modern photography modern architecture could never have been 'put across.'"[25]

However, it is still very difficult to assess the impact of the New Photography on the dissemination of Modernism. While some elements such as a preference for strong tonal contrasts and sharp focus, as well as a greater emphasis on the expression of volume, were fairly widespread, if we take a look at the collective output of professional photographers at that time, we cannot conclude that the most distinctive motifs such as extreme perspectives and abstract geometry had become the norm. It is true that a few 'iconic' images are likely to have exerted a very powerful influence, but it was undoubtedly the reproduction of photographs on the printed page that primarily affected the development of modern architecture. With the general adoption of the halftone process in the early 1890s, it had finally become economical and relatively easy to use photographs as periodical and book illustrations. The beginning of the following century saw the rise of the illustrated magazine; publications such as the German *Berliner Illustrierte Zeitung* and the French *Vue* – followed in the 1930s by *Life* in the United States and *Picture Post* in Britain – with their extreme popularity

Figure 5.3 Peter Jones department store, Sloane Square, London (1939). Architects: Slater, Crabtree and Moberly. Photographers: Dell & Wainwright. Dell & Wainwright / RIBA Collections

allowed an unprecedented dissemination of photographic images. Architectural journals in the inter-war years echoed this development, not only growing in number but also giving increasing space and prominence to photographs. This was especially true of periodicals such as the *Architectural Review* that aimed at a diversified readership, as photography was considered the most direct way to communicate architecture to a non-specialist audience. The Dutch *Wendingen* was one of the first to combine limited texts with large photographic reproductions, while the German *Das Neue Frankfurt* used photographs also on its covers and regularly published articles on photography and cinema. Both magazines, which were heavily influenced by the Neue Sachlichkeit, stopped publication in the early 1930s, but several others had been established during the previous years, including *Casabella* and *Domus* in Italy, *L'Architecture d'Aujourd'hui* in France and *de 8 en Opbouw* in the Netherlands, which were all to play a major role in the promotion of modern architecture. Other publications such as the Bauhaus books and the portfolios of *Cahiers d'Art* also extensively used photography to familiarise architects around the world with the latest architectural developments – not to mention the numerous monographs of architects published (especially in Germany) in the 1920s and 1930s, largely based on the extensive use of photographs and plans, accompanied by short texts. Perhaps the most influential publications – apart from journals – were, however, the "visual anthologies" devoted to modern architecture, from the first Bauhaus book, *Internationale Architektur*,[26] edited by Gropius (1925), to *Russland, Europa, Amerika. Ein architektonischer Querschnitt*[27] by Mendelsohn, *Die Neue Baukunst in Europa und Amerika*[28] by Bruno Taut (both 1929) and *Gli elementi dell'architettura funzionale*[29] by Alberto Sartoris (1932). In all these books, the images, especially photographs, play a much greater role than the text, in order partly to reach a wider audience beyond the architectural community and partly to demonstrate that the new architecture was a reality and not just an aspiration. As British architect H. S. Goodhart-Rendell observed in his reluctant acknowledgment of the persuasive power of the camera, "A photograph proves that a building exists or has existed; a drawing proves only that it has been proposed."[30]

Architectural exhibitions and their catalogues also relied on photography as the essential medium of communication (Fig. 5.4). As early as 1923, the exhibition *Twenty Years of British Architecture*, organised by the Architecture Club, consisted almost exclusively of photographs, which also played a crucial role in several other exhibitions and fairs such as the Deutsche Bauausstellung of 1931 in Berlin and the various Triennale exhibitions in Milan. Most famously, the exhibition on the International Style, *Modern Architecture – International Exhibition*, held at the Museum of Modern Art in New York in 1932 curated by Henry-Russell Hitchcock and Philip Johnson, was dominated by large-size photographs. The exhibition was accompanied by the equally influential book *The International Style: Architecture Since 1922*[31] and later travelled to Los Angeles, where the choice of venue, the Bullocks-Wilshire Department Store, points to the organisers' intention to advance the cause of the new architecture with the still sceptical American public. Also significant is the exhibition *Za novou architekturu* (In Praise of the New Architecture) in Prague, planned for 1938 but which instead took place in 1940 in spite of the German occupation of the city. A survey of the most significant examples of modern architecture in Czechoslovakia, its reliance on photographic images prompted avant-garde photographer František Černák to argue that its title should have been *Photography in Praise of the New Architecture*.[32]

Figure 5.4 Norwegian Pavilion, Exposition Internationale des Arts et Techniques dans la Vie
     Moderne, Paris 1937. Photographers: Dell & Wainwright. Architectural Press
     Archive / RIBA Collections

Architects had by now become aware of the expressive as well as persuasive power
of the photograph; while many of them used photographs as material for study and
analysis, or used the camera themselves as a type of visual 'note-taking', others also
wanted to be involved in the choice of images selected for publication or even in the
creation of these images. They started realising for the first time that photography
didn't just convey the 'image' of the architecture but could also convey the ideas, ideals, values and aspirations related to it. A number of regular collaborations between
architects and photographers were formed in this period. Robert Elwall argues, "At
their best such relationships became creative partnerships that led to a deeper understanding of the architecture portrayed; at their worst they smacked of censorial control
that raised fundamental questions about the role and independence of the photographer".[33] It is in fact fairly difficult – apart from very well-documented cases such as
the controlling role taken by Le Corbusier in the collaboration with his favourite
photographers – to assess the extent to which architects (and editors) influenced the
work of the photographers they employed. What we do know is that architects such
as Walter Gropius, Frank Lloyd Wright, Le Corbusier, Mies van der Rohe and Eric
Mendelsohn all took an active interest in photography and in the use of photography
to disseminate their work. Some of them deliberately promoted the repeated use of the
same images, which they considered the most appropriate to represent their buildings
and especially their ideas, thus contributing to the creation of 'iconic' images that
appeared in countless publications and still define those architects to this day.

Periodicals, books, exhibitions – not to mention advertising – therefore all contributed in the 1920s and 1930s to the dissemination of information among the architectural community and beyond at an unprecedented scale, which was essential to the development of the International Style. That this information was often mediated by the camera's peculiar vision – in the literal sense – by the photographer's individual choices, by the specific editorial or curatorial approach, may sound like an all too obvious observation, but it is one that is frequently overlooked. Too often have photographs been regarded purely as reproductions of architectural works, thus inevitably destined to fail in their attempt to replace the experience of the building. When photography's interpretative role has been acknowledged, the most recurrent criticism has been its capacity for 'deceit' – that is, to give a 'false' representation of the building. British photographer John Donat famously titled a talk given at the Royal Institute of British Architects (RIBA) in 1967 'The Camera Always Lies': a provocative intervention aimed at introducing some much-needed questioning of the accepted role of photographic representation. However, the camera cannot 'lie', simply because it cannot 'tell the truth' either. Photography does not show us what *is*, it shows us what the photographer has seen at one specific moment in time; at its best, it has the capacity of showing us what we *haven't seen*. Or, as Sean O'Hagan puts it, photography has the power to make us look at what we would otherwise overlook.[34]

The influence of the camera – an ongoing debate

So what is the impact that photography's mediation had on the process of design? Already in the inter-war period, architects and critics started asking this question. The *Architects' Journal's* editorial team maintained, "To no profession is a proper understanding of the whole creative and revelational scope of modern photography more important than our own. The architect of the near future will neglect it at his peril",[35] while the *Architectural Review* talked of the "ever-increasing debt which architecture owes to the camera",[36] probably referring to the new way in which it encouraged architects to look at volume and space. There were also notes of warning, such as the aforementioned comments by Basil Ionides, who was probably alarmed by the recurrent use of distorted views in the magazine and therefore, one can imagine, by the images' lack of 'faithfulness' to the actual buildings. The *Review*'s official photographers, Dell & Wainwright, were so influential that it was jokingly said that "the modern school of architects . . . designed their buildings not to please their clients or even themselves, but to please Dell & Wainwright."[37] Architects wanted their works to 'look good' in photographs, as they realised photography was the most effective tool of communication and publicity at their disposal. It could indeed be argued that with the advent of Modernism architecture became 'photogenic' for the first time, not surprisingly if we consider the contemporary development and reciprocal influence of the two media.

In subsequent decades, what was perceived as an idealised vision of modern architecture promulgated by photographs came increasingly under attack, especially in Britain. After all, the sun rarely shines in the British Isles, and the buildings that looked refulgent and pristine in the photographs taken shortly after their completion revealed, only a few years later, structural faults and maintenance problems. Already in 1938, a young Hugh Casson, in his review of the widely illustrated *The Modern House* by F.R.S. Yorke, said,

It is with distrust that once glances through these glossy photographs taken in strong sun, when the paint was scarcely dry upon their subjects. A suspicion that the architects are inclined to design for the benefit of other architects becomes a conviction that architecture . . . is now the handmaid of photography.[38]

Immediately after the war, a thought-provoking article appeared in the pages of the *Architectural Review* titled "Colour and Modern Architecture or the Photographic Eye".[39] Its author, painter Michael Rothenstein, acknowledges the synergy between photography and architecture – "the camera has been the great publicist of modern building"[40] – but also argues that the architect "unconsciously influenced by the camera's interpretation – is himself in danger of developing the Photographic Eye",[41] which sees in terms of black and white. Lamenting the lack of colour in the majority of contemporary British buildings, he is particularly concerned with the effect that photography might have had on "untravelled and impressionable students"[42] by fostering a conception of architecture deficient in chromatic values. However, he also expresses the optimistic opinion that, since modern architecture is now established, "the younger architects may feel free to attempt a closer exploration of texture and colour values."[43] Rothenstein was not the only one to highlight the peculiarity of photography's black-and-white vision. It has been often pointed out that numerous buildings of the International Style featured coloured surfaces, but their colours did not register in the public consciousness as they were mediated by the photographic image and therefore filtered through a black-and-white lens. The most scathing criticism of what was seen as an idealised image of modern architecture frequently presented by photography was perhaps expressed in the article 'The Craven Image' by photographer Tom Picton, published in two parts in the *Architects' Journal* in 1979.[44] Picton defines this image "an impossible dream"[45] and berates mainstream architectural photography for its sterile perfection that demands the depiction of empty spaces: "This religion of architecture is cold, excluding people from its icons".[46] It goes to the extent of comparing this flattering interpretation of buildings to the way dictators such as Stalin, Hitler and Mussolini were photographed for official pictures. Interestingly, however, and unlike many others who followed him in similar criticism, Picton does not make a connection between this type of imagery and the New Photography of the inter-war years; on the contrary, he hails that period as a high point in the relationship between architecture and photography and indirectly draws attention to their spontaneous and fruitful collaboration by observing that "architecture was integrated into cultural life. Writers, painters and photographers, well known in other areas, were interested in photography."[47] He notes that since then, by contrast, interest in and knowledge of architecture and photography has become restricted to a number of specialists. Picton also comments on the effect that contemporary "glamorous, large format photographs"[48] might have on architectural design: "Our buildings should look good in our mellow atmosphere. If they need the strong, direct sun of the Mediterranean we are building for the wrong latitude."[49] Since Picton's article, several other commentators have agreed on a criticism of commercial or mainstream architectural photography, seen as too dependent on "the imperatives of the mass media and the values they imply".[50] The indispensable role photographs play in our knowledge of architecture has been on the whole regarded as a handicap, an opinion exemplified by Kenneth Browne's bleak observation: "The thing I feel very aware of is that the camera is a splendid excuse for seeing nothing."[51]

There remains a fundamental question – did architects worldwide really try to replicate in their designs the idealised buildings they saw in the eye-catching photographs inspired by the New Photography? Or was it instead the sheer amount of images of modern buildings put in circulation by publications and exhibitions that created a lasting influence on their work?

If we look back at the nineteenth and early twentieth centuries, photographic records of existing buildings had always been essential sources for architects. The wide availability of images of historical architecture taken in all corners of the globe and avidly collected by practitioners must have undoubtedly played a major role in the development of nineteenth-century eclecticism. In addition, some of the supposedly quintessential practices of inter-war photography were just as common in preceding decades, such as excluding the human figure from the composition, photographing new buildings shortly after completion (which is still common practice) and capturing them divorced from their surroundings, which may have contributed to a tendency to design buildings indifferent to their context. Moreover, seeing the world – and buildings – in black and white had *always* been a prerogative of photographic representation. Were Bedford Lemere's sepia-toned (or even blue)[52] Victorian domestic interiors any closer to 'reality' than Dell & Wainwright's dramatically monochromatic views of modern architecture? Again, it is the fact that this world in black and white was so much more *visible* than it had been before that might have had some bearing in architects' approach to design. And even if we take this for granted, could the monochromatic vision have also been to some extent useful to architects rather than just narrowing their design tools? Photographers often express the view that black and white allows for more creativity in their craft, precisely because the resulting image is more removed from reality. As Eric de Maré put it, a

> special quality of photography is the power it allows for the imaginative formalization of the outer world into a range of grey tones between black and white . . . Every photograph is therefore to some degree an abstract, subjectively created.[53]

Contemporary photographer Charlie Waite notes that black and white encourages a process of selection in which the composition is stripped of anything superfluous; Richard Bryant explains his past use of preliminary black-and-white Polaroids before shooting buildings on colour transparency film as a device that helped him focus on composition and tonal range. The lack of colour, therefore, seems to imply for many photographers an increased attention to other elements – composition, geometry, the effect of light – which could have equally interested architects. Photography's restricted but unique vision arguably informed the design process in other ways, too, and never more so than with the advent of the New Photography, precisely because the specific qualities of the photographic medium were for the first time exploited to the full. As John Szarkowski eloquently pointed out, "The invention of photography provided a radically new picture-making process – a process based not on synthesis but on selection."[54] The inter-war period is when photographers became truly aware of this and embraced the camera's highlighting of the fragment, a prerogative that was seen not as a limitation but as an opportunity to increase understanding and insight. This is how photography has been invaluable to the architect – not because it could in any way replace the experience of visiting or using a building, but because of its potential for revealing previously unseen aspects of things and for communicating ideas. Again in Szarkowski's words, "To quote out of context is the essence of the photographer's craft";[55] therefore, expecting a photograph to tell us the whole story – tell us all about the building – is an illusory aspiration, no matter how wide the angle of the lens or how comprehensive the view. In an illuminating remark that echoes the New Vision ideals, another contemporary photographer of architecture, Patrick Reynolds, observes,

> A photograph is a sort of reduction of the world, but it is through this very reducing that something new, or something beautiful or truthful can be found. The photograph is of its subject but it is not its subject.[56]

Ultimately, it would be difficult to deny that the widespread circulation of photographic images during the 1920s and 1930s resulted in an incomplete, uneven and at times selective if not deliberately manipulated circulation of information on modern architecture; without it, however, the dissemination and exchange of new ideas in this crucial period in twentieth-century architecture would have been infinitely more limited. Its contribution to our knowledge, in particular to architects' knowledge, of Modernism cannot be underestimated. It is, however, much more difficult to evaluate the extent to which inter-war architectural photography has influenced contemporary practice. Is there a link between Modernist imagery, with its perceived emphasis on the photogenic qualities of the building, and the more recent phenomenon of the iconic building – a building that appears to be designed to be photographed? The risk of creating 'photogenic architecture' became perhaps implicit once architects started using the photograph as design tool. As Nicholas Olsberg suggests when he comments on Eric Mendelsohn's reliance on photographic studies of his Einstein Tower as a reference for his freehand sketches, "It is perhaps only a short step from such use of photography to a habit of mind that begins to conceive of a design in terms of the camera's establishing view."[57] It is a commonly accepted view – and not without reason – that Modernist photography of the 1920s and 1930s helped codify decades of professional architectural photography from the post-war years onward, and the resulting imagery constantly reaffirmed the camera's expressive power and therefore architects' aspiration to see their work validated by it. The current dominance of the photographic image in all spheres of life – much increased since the 1930s – and the speed with which these images get replaced by others seem to generate the need for highly recognisable buildings, whose complexity can be encapsulated in a single icon. Philip Ursprung observes, "There is a growing demand for highly visual architecture, for the signature building designed by a star architect to give a place, a company or political institution a certain identity."[58]

One conclusion might at this point be drawn: whenever architectural photography comes under attack, there is an underlying criticism of the architectural culture that has produced it, or even of the whole social, economic and political context.[59] This demonstrates once again how effective photography can be as a vehicle for ideas and values. An interesting and slightly dissenting perspective was offered by photojournalist Peter Baistow, who, when asked how much architecture had been affected by photography, replied, "Not as much as architects are affected by the means by which they draw it in the first place."[60] His observation reminds us of the level of abstraction that the process of design requires in the first place and the inevitable distance between any form of graphic, visual and bi-dimensional representation of a building and the building itself, with its spaces and materiality.

Notes

1 "No Steel, No Skyscraper!," *Architectural Review* 72 (1932): 219.
2 "Correspondence," *Architectural Review* 73 (1933): 48.
3 A belief that technological advances would promote social and cultural (as well as economic) advancements permeated the early twentieth-century avant-gardes, from Futurism to Constructivism to the Modern Movement. As a reaction to established ways of seeing and therefore interpreting the world, the 'new vision' could only be achieved through the use of the camera which, unlike traditional tools of visual representation, was essentially a machine. Photography – a modern, technological medium – was ideally suited to enhancing the individual's understanding of the modern world. The numerous representations of industry and its artefacts by New Vision photographers reflect this faith in technologies. On the theme of technology and the New Vision, see Eleanor M. Hight, *Picturing Modernism: Moholy-Nagy and photography in Weimar Germany* (Cambridge, MA: MIT Press, 1995) and Maria Morris Hambourg and Christopher Phillips, *The New Vision: Photography between the World Wars* (New York: Metropolitan Museum of Art / Harry N. Abrams, 1989).
4 László Moholy-Nagy, *Painting, Photography, Film* (London: Lund Humphries, 1969), 7. This is the first English edition of the 1925–27 original.
5 As cameras became smaller, and therefore the size of the image format decreased, sharper lenses were needed to retain definition. The early twentieth century also saw the introduction of the astigmatic lens, which corrected distortions and therefore avoided a lack of sharpness in certain areas of the picture.
6 As early as 1919, the *Architectural Review* had published an article on the possible effects that aerial photography might have on the visual representation of architecture. The author observes, "After all, an aerial photograph is rather similar to a paradox, in that it sees truth from a singular angle and, as such, expounds it; and who will traverse the statement that an edifice seen from a singular angle is often more captivating than when seen from an orthodox standpoint?" (Capt. Gordon H. G. Holt, "Architecture and Aerial Photography," *Architectural Review* 45 (1919): 9). The novelty of seeing the built environment from above and from a considerable distance for the first time inspired photographers to look for less obvious viewpoints. The resulting bird's-eye view (together with the opposite worm's-eye view) emphatically expressed the height of tall structures such as skyscrapers, pylons and factory chimneys – symbols of modernity in architecture. Aerial views of groups of buildings also highlighted geometry and pattern.
7 These techniques highlighted the effectiveness of juxtaposition as a way of making a statement, therefore encouraging photographers to create contrasts and comparisons between the different elements of their compositions.
8 Quoted in Peter Galassi, "Rodchenko and Photography's Revolution," in *Aleksandr Rodchenko*, ed. Magdalena Dabrowski, Leah Dickerman, and Peter Galassi (New York: Museum of Modern Art, 1998), 122.
9 Alexander Rodchenko, "Puti sovremennoi fotografii (The Paths of Modern Photography)", *Novyi LEF* 9 (1928): 31–39. Quoted in Selim O. Khan-Magomedov, *Rodchenko: The Complete Work* (London: Thames and Hudson, 1986), 225.
10 Franz Roh and Jan Tschichold, *Foto-Auge: 76 Fotos der Zeit* (Stuttgart: F. Wedekind, 1929).
11 In 1920 Sheeler collaborated with Paul Strand on the short documentary film *Manhatta*, which presented a day in the life of Lower Manhattan.
12 Robert Elwall, *Building with Light: The International History of Architectural Photography* (London: Merrell, 2004), 122.
13 Eric Mendelsohn, *Amerika: Bilderbuch eines Architekten* (Berlin: Mosse, 1928).
14 Quoted in Beaumont Newhall, *The History of Photography* (New York: Museum of Modern Art, 1937), 11.

15 Richard Pare, *Photography and Architecture, 1839–1939* (Montreal: Canadian Centre for Architecture, 1982), 25.
16 Jaroslav Anděl, *The New Vision for the New Architecture: Czechoslovakia, 1918–1938* (Zurich: Scalo, 2006), 9.
17 Jaromír Krejcar, trans., *Život; Sborník nové krásy* (Prague, 1922).
18 Anděl, *The New Vision*, 10.
19 Krejcar, trans., *Život*, quoted in Anděl, *The New Vision*, 10.
20 Karel Teige, "Foto Kino Film," in Krejcar, trans., *Život*, quoted in Anděl, *The New Vision*, 10.
21 Cervin Robinson and Joel Herschman, *Architecture Transformed: A History of the Photography of Buildings from 1839 to the Present* (Cambridge, MA: MIT Press, 1987).
22 Elwall, *Building with Light*, 122. Robinson and Herschman, *Architecture Transformed*, 110.
23 Robert Elwall, *Photography Takes Command: The Camera and British Architecture, 1890–1939* (London: RIBA Heinz Gallery, 1994), 77.
24 Philip Morton Shand, "New Eyes for Old," *Architectural Review* 75 (1934): 12.
25 Ibid.
26 Walter Gropius, trans., *Internationale Architecktur* (Munich: Albert Langen, 1925).
27 Eric Mendelsohn, *Russland, Europa, Amerika: Ein architektonischer Querschnitt* (Berlin: Rudolf Mosse, 1929).
28 Bruno Taut, *Die Neue Baukunst in Europa und Amerika* (Stuttgart: Julius Hoffmann, 1929).
29 Alberto Sartoris, *Gli elementi dell'architettura funzionale* (Milan: Hoepli, 1932).
30 H. S. Goodhart-Rendell, "Architectural Exhibitions," *Architect and Building News* 150 (1937): 33.
31 Henry-Russell Hitchcock and Philip Johnson, *The International Style: Architecture since 1922* (New York: Norton, 1932).
32 František Černák, "Největší výstava fotografických zvětšenin," *Foto-Noviny* 21 (1940): 18–20. Quoted in Anděl, *The New Vision*.
33 Elwall, *Building with Light*, 128.
34 Sean O'Hagan, "Look This Way," *The Observer*, October 16, 2005, http://www.theguardian.com/theobserver/2005/oct/16/art
35 "The Film Institute and Architecture," *Architects' Journal* 78 (1933): 375.
36 "Photography," *Architectural Review* 77 (1935): caption to plate iv.
37 "Notes & Topics," *Architects' Journal* 103 (1946): 445.
38 Hugh Casson, "Architecture at Home," *Architectural Review* 83 (1938): 90.
39 Michael Rothenstein, "Colour and Modern Architecture or the Photographic Eye," *Architectural Review* 99 (1946): 159–163.
40 Ibid. 159.
41 Ibid.
42 Ibid.
43 Ibid. 163.
44 Tom Picton, "The Craven Image: Or the Apotheosis of the Architectural Photograph," *Architects' Journal* 170 (1979): 175–190, 225–242.
45 Ibid. 176.
46 Ibid.
47 Ibid. 186.
48 Ibid. 176.
49 Ibid. 184.
50 Kenneth Frampton, "A Note on Photography and Its Influence on Architecture," *Perspecta* 22 (1986): 41.
51 Picton, "The Craven Image," 242.
52 Bedford Lemere made frequent use of the cyanotype, a photographic printing process that produces a cyan-blue print. In order to understand their process of image-making, it should also be borne in mind that they carefully rearranged furniture and objects before photographing interiors.
53 Eric de Maré, *Photography and Architecture* (London: Architectural Press, 1961), 21.
54 John Szarkowski, *The Photographer's Eye* (New York: Museum of Modern Art, 1966), 6.

55 Ibid. 70.
56 Quoted in Patrick Reynolds, "The Eyes of the Beholders," *Architecture New Zealand* 3 (2005): 114.
57 Nicholas Olsberg, "The Shattered Glass," *Architectural Review* 235 (2014): 75.
58 Philip Ursprung, "Built Images: Performing the City," in *Images: A Picture Book of Architecture*, ed. Philip Ursprung, Ilka Ruby and Andreas Ruby (Munich: Prestel, 2004), 4.
59 Tom Picton was fairly explicit on this point: "Arid and soulless and pompous photographs too often, on the evidence of the buildings, accurately portrait an arid and soulless and pompous profession" (Picton, "The Craven Image," 190).
60 Picton, "The Craven Image," 241.

Bibliography

Anděl, Jaroslav. *The New Vision for the New Architecture: Czechoslovakia, 1918–1938*. Zurich: Scalo, 2006.

Bergera, Iñaki, trans. *Fotografía y arquitectura moderna en España / Photography & Modern Architecture in Spain, 1925–1965*. Madrid: Museo ICO / La Fabrica, 2014.

Brockhaus, Christoph. *Werner Mantz: Architekturphotographie in Köln 1926–1932*. Koln: Locher, 1982.

De Maré, Eric. *Photography and Architecture*. London: Architectural Press, 1961.

Dufek, Antonín. *Jaromír Funke: Between Construction and Emotion*. Brno: Moravian Gallery / KANT, 2013.

Elwall, Robert. *Building with Light: The International History of Architectural Photography*. London: Merrell, 2004.

Elwall, Robert. *Photography Takes Command: The Camera and British Architecture, 1890–1939*. London: RIBA Heinz Gallery, 1994.

Elwall, Robert, and Valeria Carullo. *Framing Modernism: Architecture and Photography in Italy, 1926–1965*. London: Estorick Foundation, 2009.

Fanelli, Giovanni. *Storia della fotografia di architettura*. Bari: Laterza & Figli, 2009.

"The Film Institute and Architecture." *Architects' Journal* 78 (1933).

Hambourg, Maria Morris, and Christopher Phillips. *The New Vision: Photography between the World Wars*. New York: Metropolitan Museum of Art / Harry N. Abrams, 1989.

Heckert, Virginia, trans. *Albert Renger-Patzsch: History of Photography* 21 (1997): 177–259.

Higgott, Andrew, and Timothy Wray, trans. *Camera Constructs: Photography, Architecture and the Modern City*. Farnham: Ashgate, 2012.

Hight, Eleanor M. *Picturing Modernism: Moholy-Nagy and Photography in Weimar Germany*. Cambridge, MA: MIT Press, 1995.

Hiss, Tony. *Building Images: Seventy Years of Photography at Hedrich Blessing*. San Francisco: Chronicle Books, 2000.

Ibolya, Csengel-Plank, Hajdú Virág, and Ritoók Pál. *Fény és Forma: Modern Építészet es Fotó, 1927–1950 / Light and Form: Modern Architecture and Photography, 1927–1950*. Budapest: National Office of Cultural Heritage, 2003.

Mazza, Barbara. *Le Corbusier e la fotografia: la vérité blanche*. Florence, Italy: Firenze University Press, 2002.

Mendelsohn, Eric. *Amerika: Bilderbuch eines Architekten*. Berlin: Mosse, 1928.

Moholy-Nagy, László. *Painting, Photography, Film*. London: Lund Humphries, 1969.

Newhall, Beaumont. *The History of Photography*. New York: Museum of Modern Art, 1937.

Olsberg, Nicholas. "The Shattered Glass." *Architectural Review* 235 (2014): 71–79.

Pare, Richard. *Photography and Architecture, 1839–1939*. Montreal: Canadian Centre for Architecture, 1982.

Picton, Tom. "The Craven Image: Or the Apotheosis of the Architectural Photograph." *Architects' Journal* 170 (1979): 175–190, 225–242.

"Photography." *Architectural Review* 77 (1935).

Robinson, Cervin, and Joel Herschman. *Architecture Transformed: A History of the Photography of Buildings from 1839 to the Present*. Cambridge, MA: MIT Press, 1987.

Rothenstein, Michael. "Colour and Modern Architecture or the Photographic Eye." *Architectural Review* 99 (1946): 159–163.

Shand, Philip Morton. "New Eyes for Old." *Architectural Review* 75 (1934): 11–12.

Stebbins, Theodore E., Gilles Mora, and Karen E. Haas. *The Photography of Charles Sheeler: American Modernist*. Boston: Bullfinch Press, 2002.

Szarkowski, John. *The Photographer's Eye*. New York: Museum of Modern Art, 1966.

Ursprung, Philip, Ilka Ruby, and Andreas Ruby. *Images: A Picture Book of Architecture*. Munich: Prestel, 2004.

Walker, Paul, and Justine Clark. "The Architectural Photograph: Image and Ideology." *Architecture New Zealand* 3 (2005): 108–110.

Zimmerman, Claire. *Photographic Architecture in the Twentieth Century*. Minneapolis: University of Minnesota Press, 2014.

6 Photography and architecture

From technical vision to art and phenomenological (re)vision

Iñaki Bergera

Introduction

Photography is a visual construct – an autonomous abstraction of reality intimately connected to what it portrays. The photography of architecture is constructed, in a similar way to the building itself, as it manipulates indistinctively light, form, matter and space. It overlaps metonymically image and language, the architecture of the photograph with the photography of architecture. Photography is effectively a technique of architectural visualization that represents and documents the spatial, visual and material values of architecture whilst contributing its own distinct aesthetic principles that have a significance of their own. As it captures the space and renders it in two dimensions, the position and angle of the camera will define a kind of crystallization of that space. Therefore, in the photography of architecture, the object can end up being an 'excuse', something that becomes absolutely transfigured and subjected to a completely autonomous rhetorical discourse at the margin of the building itself.

This phenomenon was first manifest in the early twentieth century when, through photography, the modern gaze found in the architectural avant-garde the perfect pretext for the construction of its own new and objective visual identity. The same also happened in reverse: rationalist architecture embraced the photographer's new language in order to disseminate its particular messianic architectural-aesthetic language. This union consolidated and consecrated in the twenties and thirties the new architecture – rapidly transformed in style by the iconic dissemination of its images – and defines and articulates the relationship between architecture and photography even today. These two languages are then inseparable. Despite the great theoretical and interpretative potential of this symbiosis between architecture and photography, until recently its analysis had not attracted the attention and recognition it deserved from the critics and historians of both disciplines. This lack of attention may well be one of the reasons: the potential of photography to inform architectural discourse and practice in positive ways has – by and large – been underutilised.

In addressing one particular aspect of this scenario, Robert Elwall, the RIBA's former curator of architectural photography, stated in 2004,

> Architectural historians too often treat photographs as if they were the buildings themselves and not particular interpretations of them made at particular moments. For their part, photographic historians [. . .] continue to view architectural photography as an arcane technical subject, a murky cul-de-sac deviating from photography's true path towards the status of art.[1]

This positive, progressive and ever more specific attention on the relations of this union opens a discourse on interests which are not strictly disciplinary or historiographical. When we consider photography from its technical perspective, and its influences on the architectural practice, it can claim to be an inherent component of modern architecture but this fails to fully explore its potential, which, as Elwall hints at, is more fruitfully explored and understood by opening the discourse we have about it to the influence of art practices.

The photography of architecture has been, and still attempts to be, a tool that examines and documents architecture. Furthermore, architecture and the urban landscape can be said to have been two of the greatest inspirations for artists who have for over a century described the complexity and uncertainty of 'modern' societies. Once you strip the mask of photography away from architecture, and assume the impossibility of a faithful portrait of the architectural object – the medium's last refuge of 'usefulness' – what opens up is a useful and revealing consideration of photography as an art and social practice. After surpassing figurative and realist painting as the most faithful method of representing the constructed landscape in the late nineteenth century, photography was endowed with a seemingly inescapable documentary capacity. However, in reality, photography, the technique of architectural visualization par excellence actually influences and determines the very design of architecture and, to date, this influence has been full of problems and negative consequences for architecture itself.

This has been understood, if understated, and architects have been captivated by the power of photography as an autonomous representation of the architectural object for at least a century. Indeed, with the complicity of the photographers, they can be said to have perverted the character of their discipline in favour of an obsessive search for certain 'image types' that bestow on their project a particular autonomous aura of visual materiality. At its worst, the building's facade becomes its own super-imposed mask that seeks to project the face of its photographic portrait so as to ensure the building ends up in turn on the cover of an architecture magazine (Figure 6.1). In the process, not only has architecture been perverted, but architectural photography has ended up suffering certain degradation: first through its promotional misuse and subsequently through its vast diffusion through the mass media – initially in print in the twentieth century and more commonly today through the uncontrolled and alienating circulation of images on the Internet.

In this context of the oversupply of architectural imagery today, we witness in addition to a deliberate aesthetic manipulation of the image the anesthetization of architects, users and public opinion to such an extent that they now often fail to consider buildings from a human or functional perspective – a criticism also levelled in a different tone at the effects of photography in the early twentieth century. The resulting new 'photographic narrative' of architecture, therefore, lacks credibility. Its failure to prioritize the inherent human nature of the spaces we inhabit has actually had significantly negative effects for the quality of contemporary architecture. To tackle this issue – and venturing to guess possible solutions – this chapter gives a discursive overview of the contemporary theoretical, cultural and artistic context in which architectural photography operates and suggests that certain artistic tendencies already evident in architectural photography offer a way out of the negative set of influences that too often in the history of photography have manifest themselves in architectural representation and production.

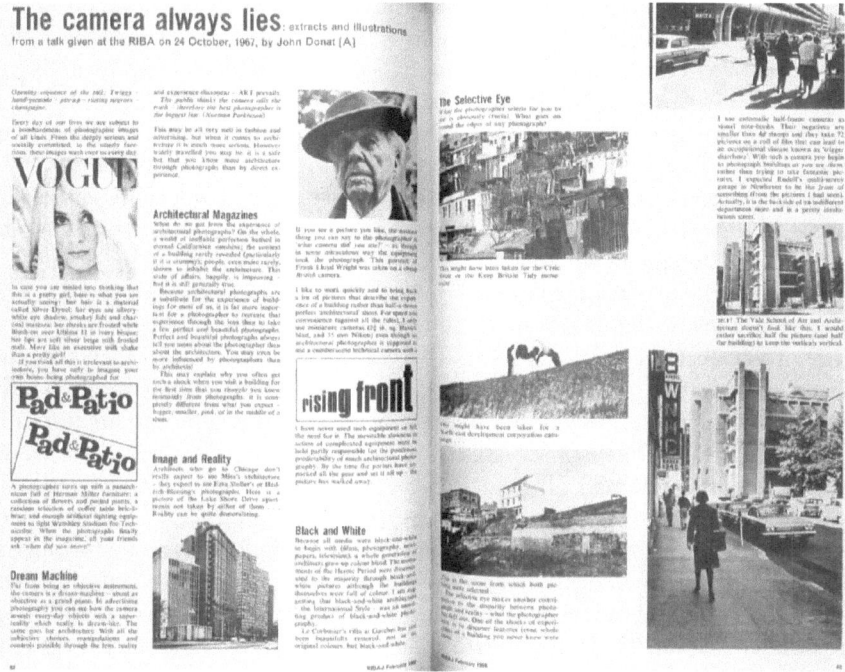

Figure 6.1 Double page of the article "The camera always lies" by John Donat, published in the *RIBA Journal 5* (1968) and lectured on October 24, 1967: a critical visual statement to the use of architectural photography

From icon object to privileging the human

In the early twentieth century, the artistic avant-garde, in architecture and photography, had to respond to the transformation initiated by the specialized publications. The building could no longer rely upon a single image;[2] a narrative sequence was required that led to collage and a graphical montage of photographs in juxtaposition with the text, which was intended not just to illustrate the article but also to orchestrate a message. It was through this set of techniques that the transformation of architecture into an object of consumption, as Beatriz Colomina pointed out,[3] was manifest. In this analysis, the overlapping of virtual images from an architectural project with commercial photographs of architecture in which both types shared in the construction of an idealized and aestheticizing sophisticated imagery distorted the discipline of architectural photography, which until then had been more 'innocently' documentary. In the worst cases, photographers then, as now, became architects of the image by *making* photos rather than *taking* them. Architects also became, literally, the photographers of the constructed world who, on occasions, even saw the construction of a particular photographic image as more important than the tectonic, material and functional values of their architecture itself.

Today, photographs of architecture are intended for a certain public and are designed for a specific use and context. In the past, this was for specific specialized magazines, but today it entails an uncontrolled diffusion of images through web pages, blogs and social networks. Whether of real or proposed projects, it is arguable that these images have become a banal masked ball for the endogamous use of a baseless architectural criticism. To understand and analyse it, knowing what and who is involved in the photographic production, who is actually behind the commission and pays for it, is essential.[4] Without a knowledge of this background, for example, we cannot understand how and why the so-called new objectivity photography of the early twentieth century essentially ended up 'iconizing' architecture. Nor can we understand how some buildings that appear impressive on a page fail to live up to expectations in reality. In this regard, the words of Alberto Campo Baeza are revealing: "In the end we architects depend on the photographers. I've always maintained that a bad architect with a good photographer is a hypocrite, and a good architect with a bad photographer is an imbecile."[5]

Photography then plays a role in architecture that is fundamental and, at times, fundamentally distorting. It is something that has only been increased in recent years because of the recent economic crisis. In this context, the ease with which we generate rendered architectural projects, and the lack of opportunities to actually build, has turned the digital and photographic undertaking of architecture into an end in itself. The avant-garde and futurist processes of modernity, and the more speculative visual narratives of post-modernity, all used drawing, collage and the photo montage to configure of its own architectural corpus. Today, the flow of both real and virtual images has become the tool with which architectural praxis is building its own formulations.

One key difference between the current scenario and what happened in the golden age of the symbiosis between buildings and their photographic representation in the early twentieth century is the popularization of photography in quantitative and qualitative terms. The explosion of digital photography has brought the professional techniques of the photographic closer to any well-meaning amateur enthusiast. Not so long ago, these techniques were in the exclusive remit of those operating an optical bench camera. They were the exclusive preserve and hallmark of the specialists of 'architectural photography'. For example, the skill, expertise and mastery embodied in the pictures of Julius Shulman (Figure 6.2) or Ezra Stoller are not comparable with the best photographs made today with digital sensors and professional lenses.

The contemporary world lauds a visual culture that privileges the power of the image over other sources of knowledge, thus facilitating the saturation of visual stimulation to which we are subjected. Digital photography and the Internet are fundamental technological players in this context both in terms of its democratizing of photographic techniques and in the dissemination of the results. This means that images of architecture are now more prevalent than at any time in history. Today, the architecture that the promiscuous camera and forms of dissemination it has at its disposal prefers is photogenic and iconic. In its worst manifestations, the camera turns architecture into a kind of pornographic object that satisfies the hedonistic gaze of the author but gives only a fleeting and superficial experience to the keen observer. The seductive quality of the image prevails over the simple information that it conveys.

However, this is not just a question of production and dissemination. For Herzog & de Meuron, "the appearance of a building is in the eye of the beholder", and given the power of the viewer in this formulation, the photographer takes up the role of mediator and propagator. In this capacity it is the interpretation of the final recipient of the image that matters and, in a commercial context, this often means consumers, readers of images in architectural magazines and websites. For these readers, it is assumed that the object presented and the space that it contains must be, above all else, photogenic. It must, therefore, also be open to any necessary visual manipulation the saturated universe of architectural

Figure 6.2 Cover of the book *Photographing Architecture and Interiors* by Julius Shulman (New York: Whitney Museum of American Art, 1962). The book condenses the master's technical proficiency and expertise on the field

marketing demands. Often, this means it must be free of the human traces of use. As a result, these images of architecture represent a parallel existence for buildings.

Considered in this way, it is also arguable that – paraphrasing Marshall McLuhan[6] – the medium of architectural photography has become the message of architecture itself. In both the period of early modernism and today, this message was and is one of the object status of architecture communicated massively through modern modes of visual communication. Today, a subtle difference has emerged in that the message is multiply reproduced by a seemingly infinite number of photographers. In particular, in this scenario, the power of the image – and any thoughtful meaning it seeks to communicate – has been lost. As Joan Fontcuberta, a lucid commentator on the state of contemporary photography explains, "We are surrounded by the capitalism of images and excess that, more than just plunging us into the suffocating world of consumption, confronts us with the political capability of dismissing, reducing, or censuring them".

However, Fontcuberta does not limit this analysis to a hopeless criticism but goes on to suggest ways out of this dead end:

> Therefore, two possible strategies arise: to propose a kind of visual ecology that delimits the unconstrained production of images, encouraging the recycling of existing images; or to not concentrate on the images that proliferate, but rather on those that are lacking.[7]

If we focus on the more intellectually stimulating of these two tasks and ask what architectural photographs are lacking, an answer can be found in the poetic, phenomenological and intellectualized ways of looking offered by the artist that opens architecture to a more complex, coherent and multi-faceted reality and which tackles the unavoidable human, urban and social dimensions of the interdisciplinary architectural world.

The architectural photography that promoted the modern movement was one that simply contemplated the architectural subject as an object. As a consequence, the architecture of the period was shaped accordingly and left aside the human identity of the living space; the repeated practice of photographing a building empty, without furniture, plants or people ended up dividing the 'author' of architecture from those who inhabited the designed building as an everyday life experience. Indeed, it could be argued that this type of architectural photography is a form of 'normative' and indicative of the latent anxiety between the reality on the ground and the reality the image aspires to capture.[8]

In contrast to this, one can highlight the photographs of the work of Herman Hertzberger who, for decades, has insisted his buildings be photographed in use – with inhabitants using spaces as they, not the architect, intended. Similarly, the praxis of the architectural photographer Iwan Baan embodies this more phenomenologically. Doing without many of the technical supports of photography such as a tripod, Baan's photographs focus on telling human stories that have architecture as their spatial and urban setting. In his work, we see an architectural photography that has come down from its ivory tower – of technique, exclusive professionalism and object-icon focus – to privilege the human.

In search of the essential portrait

Since its very beginning, the strictly documentary work inherent in the photographic recording of architecture had the honest desire of celebrating architectural beauty in order to perpetuate it, but it promptly became devalued by blurring the credibility of the image. "Photographs of architecture are forms of portraiture", writes Zaha Hadid, "constructing a parallel and often more lasting memory and identity of the building as a living (or sometimes dying) organism in itself."[9] If when we refer to the photographic portrait we submit the image to a comparison of identities – the subject that the image presents and the person that is hidden behind it – we are able to do the same with architectural photography. In this scenario, we are capable of differentiating the image we see from what it really represents. As with a portrait – whether painted or photographed – the building in the architectural photograph tends to be shown in its best profile with its deficiencies being hidden. Hence the more idealized the portrait, the more surprising is the real physical experience. At times, it can appear that the building has been unmasked once seen in reality.

Faced with the sensory experience of the architectural reality, a circumspect photographic vision will always accept its partiality. This is the case despite repeated and dangerous attempts to use the technologies of photography to their full extent, such as wide-angle lenses and low-level perspectives that attempt to capture the greatest amount of space possible, but very clearly distort human vision. Architectural photography, then, does not substitute or supplant the phenomenological and anthropological reality of the constructed world. The photographic gaze is never, and cannot be, innocent. As a medium, photography inevitably involves a process of discernment and selection of the object it observes, and in each act of looking, there is an expectation of meaning. Considered in this way, the reality of the photography of architecture, as we understand and discuss it today, involves an acceptance that the building represented is not the primary target and outcome of the photograph.

On the contrary, the building-subject of architectural photography is simply the formal and visual substance through which the observer generates an evocative response. It is not so much a question of *what* we see but rather *how* we see. Our cognitive processes are more important and override any simple and passive act of contemplation. Photography, then, must not be seen as a simple technique of architectural visualization but also a frame for the semiotic and phenomenological interpretation of the architecture as an image of itself. If we see it in this way, we can begin to use photography differently – perhaps even to intimate and express those unquantifiable aspects of architecture, to reveal the "existential space" Norberg-Schulz described.[10] Understood and used in this way, photography can be seen as a perceptive existential dimension of man that receives its essence from the location it presents. In this regard, Michael Leonard argues, "The final product in the perceptual process is a single sensation, a 'feeling' about that particular place."[11] Likewise, photography is capable of producing a synaesthesia of sensations and of evoking different perceptive reactions for each spectator. Addressed in this way, the appropriate phenomenological visualization of the environment and the 'atmosphere' of location could improve the quality and influence of architectural photography. Key to this is the fact that the 'atmosphere' we hint at here is certainly the least objective and most intangible aspect of architecture and is thus unequivocally linked to personal interpretation – of the photographer and the person who contemplates the image.

If we broaden the content and contemporary scope of architectural photography in this way, it has influence and consequences for architecture. To achieve this, however, it is useful to distinguish clearly between a photograph which is *prêt-à-porter* from one which is *haute couture* – between a casual architectural photograph and a professional one. Each one has its own creators and receivers, the difference being marked by the intellectualization of the photographer and the recipient's gaze, as explored by 'Fields of Vision' (Figure 6.3) by Herbert Bayer.

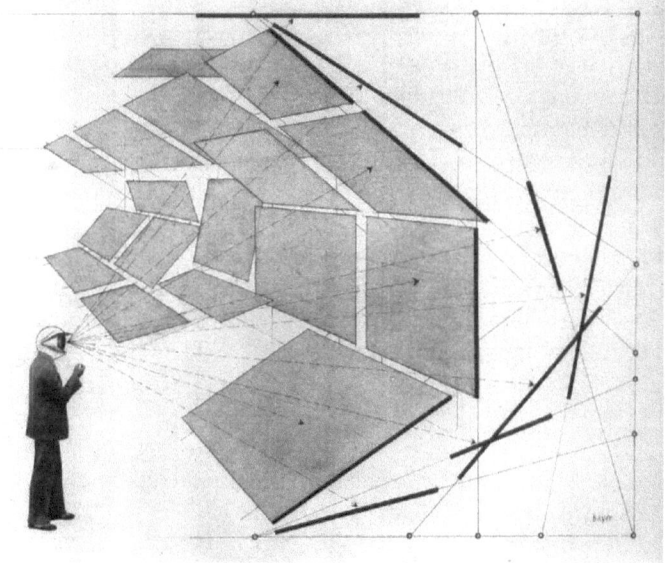

Figure 6.3 'Fields of Vision', a diagram by Hebert Bayer for the 1930 catalogue of Deutscher Werkbund, exemplifies graphically the non-lineal and fragmentary nature of vision

Created by the artist-photographer for the gallery space, the professional image is open to different interpretations and justifications. That said, for many artist-photographers, architecture and the urban landscape are pretexts for their personal practice and their framed pictures may be read as 'displayed commodities' in a similar way that the glossy professional and commercial shots are published in magazines and books. Therefore, within the artistic world, we also have to distinguish between photography conceived by and connected to the markets of the cultural industries and that which is conceived, at least at inception, as a more theoretical, open laboratory of ideas and speculation. Faced with the de-professionalization of architectural photography, as it had been understood and practised until recently, it is arguable that the more multi-faceted work of visual artists is where architectural photography can find its freest form.[12]

In contrast to the work of artists exploring the potentialities of the photographic medium to the full, commercial photography tends to *portray* architecture as an isolated formal object free from human constituent, whereas artistic practice *dialogues* with human narratives.[13] This artistic conversation is evident a home in the neutral, non-aestheticized domains represented in the work of Andreas Gursky, Thomas Ruff and Candida Höfer; their supra-realist architecture of urban landscapes and the architecture they magnify through the hyper-visuality of scale being perfect commentaries on this scenario. It is also evident in the work of artists such as James Welling and Jose Dávila, who, exhausted by the exalted and iconic modern movement architecture portrayed by the masters Shulman, Stoller and Hervé, subjected it to a deliberate and evocative visual process of figurative decomposition. Similarly, Hiroshi Sugimoto offers an unfocussed and intangible portrait of the desire to break free of literal and iconic visions of architecture and its image. In his works, we see a search for the mentally subjective and evocative.

As the work of these artists suggests, the relationship between art and life takes many different paths, and the twentieth century's avant-garde attempt to 'artify' reality was only one. With the advent of postmodernism in the 1970s, this ratified reality took a very different path of introducing human experiences into inbred artistic discourses. Experience became the pretext for the work of art, displacing the understanding of art as object with that of art as idea – or art as poetics. The photographer of architecture today should be more a witness than a narrator of what he or she chooses to portray. We are witnesses because, being present, we record. But at the same time, we operate at a distance, which somehow absolves us, as spectators, of any responsibility towards what the photograph reveals. Art emerges from this blurred frontier between the visual and the interpretive as an expression of an interpersonal agreement between the photographer, the final observer of the image, and the use that is made of it, especially by the media. Alien to this potentiality, however, architectural photography – as a technology of vision – remains almost exclusively focused on depicting the object and not the narratives and human stories it includes.

Art-photography in search of meaning

The critical, ethical and aesthetic capacity that has enabled, and will enable, architectural photography to last exists exclusively in the choices made by the photographer. As Eric de Maré wrote,

> The photographer is perhaps the best architectural critic, for by felicitous framing and selection he can communicate direct and powerful comments both in praise

and protest. He can also discover and reveal architecture where none was intended by creating abstract compositions of an architectonic quality.[14]

A well-intentioned photographer has the ability to resolve the tension between cognitive representation and perception, between the real and the represented, between image and language. "A camera can only deal with the visible", wrote the American photographer Stephen Shore. "A photographer trying to communicate his or her perception of the currents below the surface of things has to find instances where these currents are visibly manifest. There is artistic freedom then, but it is one premised on reality."

In alignment with the aforementioned comments is the old Arab saying, "The apparent is the bridge to the real." For many photographers, architecture certainly serves this function,[15] as John Berger indicates when he says we photograph to give authenticity to a set of appearances. Photographs, Berger suggests,[16] do not translate reality but rather *cite* it. John Szarkowski puts it differently. For him, a photograph "can be about a building but it cannot be one".[17] Consider in this context, then, that the progressive abstraction and conceptualization of architectural photography has resulted in it being submitted to an excessive codification far removed from the practice of architecture. In architectural photography, buildings end up acquiring the condition of *ready mades* – in the expectation of a poetic reaction. Therefore, the refrain of the artist Barbara Visser, "What you see depends on what you are looking for", seems more apt than that of Frank Stella: "What you see is what you see". In the context of architectural photography, then, it is up to the photographer to take part in the mediation process to facilitate or interfere in the search for meaning.

While referencing reality, the photography of architecture is more than a document. Indeed, it is arguable that it is an associative and romantic evocation of memory, desire and imagination in search of *pathos*. In this sense, the frame becomes essential and defining, as Robert Campbell pointed out: "Through the view-finder you place a frame around something, isolating it and giving it importance while supressing everything that remains outside of it. The photograph, in other words, involves in addition to everything else, the elimination of context."[18] This can be seen in the fragmentary nature of the representation that eliminates and classifies reality, underlining the arbitrariness of the possible meanings that can be derived from it. Each photograph is a kind of synecdoche, a metaphoric recoding of reality as a fragment.

These contradictions between the documentary, the descriptive, the artificial and the evocative allow space for a form of architectural photography that not only allows us to see architecture but also facilitates our perception and internalization of it. A good example of this is found in Walter Niedermayr's examination of the work of Sejima and Nishizawa. The synergy in the work of SANAA is evidenced fully in capturing the natural light and the ephemeral nature of the spaces. It is a question of capturing essential qualities, whether visual or phenomenological, together. In arguing for this, however, Niedermayr is arguing from his position as an artist himself, but not all professional photographers are artists and not all artists are capable of reading an architectonic space fully.

One artist who certainly is capable of doing this is Hans Danuser, who was asked at the end of the 1980s to photograph the chapel of Sogn Benedetg in the Alps by Peter Zumthor. The atmospheric monochrome photographs Danuser presented were far removed from the anonymity and coldness of the direct photograph typical of the time. They favoured an explicit artistic and inherently human interpretation of the space that also reinforced the identifying hallmarks of Swiss art and architecture. It was, according to Phillip Ursprung, a performative and sequential work that highlighted the compactness of the building through a succession of image fragments rather than a reading of the whole.[19] As fragments, they refer to the human experience – that of the photographer and the user. They thus evoke actually living, touching and facing the building as an overall human experience.

Since the 1990s, Peter Zumthor has worked with Hélène Binet, a photographer who has also worked with Zaha Hadid, among others. Her photographs portray the general from the particular – understanding architectural photography as the communication of a strong but slight experience. It is always many sided and rejects any attempt to recount a single truth about the building,

> whose complexity can never be caught in one global image. Buildings are a collection of moments experienced. The images need each other to create an echo, to escape the static nature of singular, global observations, and to generate virtual spaces.[20]

For Daniel Libeskind, Binet's photography "reveals the interior intensity of architecture, materialising the nature of the light, texture and density, within a totally conceptual composition" (Figure 6.4).

Figure 6.4 Interior view of Therme Vals, the spa complex built by the Swiss architect Peter Zumthor. Photograph by Hélène Binet. © Hélène Binet

This is not new and it is new. Etymologically, 'to photograph' is to draw with light, crystallizing it on the textures and forms that define the architectonic space. For Binet, light constructs the evanescent atmosphere of the architecture in relation to time. Light requires shade as music requires silence. Her photographs record the flow of mutual dependencies that is staged in architecture. Binet is thus more narrator than photographer in the way that she emphasizes what she tells rather than how she tells it. Whether focusing on 'the what' or 'the how', the object is the same: architecture. The disjuncture that appears between the work of these artists and the architectural photography that proliferates today is not so much between the personal gaze as emotional and subjective, and the literal rhetorical look as alienating. On the contrary, it is between labelling the photograph as artistic or professional.

In some way, any artistic gaze is essentially and strictly professional, as it is used to portray the part with the whole elegantly. No good professional photographer would give up the search for the soul of a project in favour of its mere formal and aesthetic visualization. However, it is artists who more readily imbue their work with a certain aura and a latent human dependability that can act as an antidote to the conventional hyper-realism of the gaze and the pre-fabricated saturation of architectonic meanings. If we consider again the photographic codification suggested in 1978 by the MoMA curator of photography, John Szarkowski,[21] good architectural photography should aspire to overlap its condition of mirror and window, describing space through a deliberate but progressively more transparent gaze – a gave that cannot overlook human referentiality in architecture.

Conclusion

Ethically, architecture and photography both need and exclude each other. The worst architecture can be beautifully photographed. Equally, it is impossible to coherently represent certain spaces and buildings in images. However, when good architecture encounters an appropriate, consistent and suggestive photographic representation, it inspires us to experience that space. Good and personified architectural photography becomes a catalyst for hopes and yearnings that re-nourishes the creativity of the architects, and it should not be forgotten that photography has also applied a positive influence in architecture. Indeed, given the sense of unease produced by the growing visual saturation, it seems right to remind ourselves and even foster a positive discourse about the relationship between photography and architecture.

This positive discourse is even more pressing given that, for better or worse, we live in a world saturated with architectural images that we consume voraciously, without any possibility of digesting them, or even less of savouring them. This fact has produced a sort of visual agnosia that seems to incapacitate the comprehensive and critical analysis of what we see. Far from making our understanding clearer, this saturation makes it even more opaque, as evidenced by much of the trivial contemporary architectural criticism in evidence today. Without sense or discernment, images produce an intellectual blindness and necrosis of the memory.

In contrast to this, we need to develop a discerning and sensitive visual memory that is capable of utilizing architectural photography as a source of non-verbal knowledge and an aid to thought. In this light, Helio Piñón argues that architectural photography – not so much in itself, but as an analytic tool for architects, historians, critics and teachers of architecture – can through intellection and visual thought open a path to genuine aesthetic judgement. For this to happen, we have to see photography as a tool and product of visual thought and better understand how this structures our judgement of buildings. In contrast to the meagre multiple presentation of distorted, easily consumed images, this 'tool' has to have a 'human consistency' ingrained with anthropological values and social discernment.

This is something not just applicable to photography, however. Architecture itself, needs to focus on the same set of human values. It is a two-way street. This realization of a mutual symbiosis is evident in comments from Robert Elwall, who urged us to read the evolution of architecture and photography as the evolution of a union of the disciplines. In contrast to reading the relationship of photography and architecture as one of natural coalescence of independent artistic practices, he suggests a more complex set of motives.

> If, therefore, we look at the development of architectural photography less from the perspective of art appreciation and more from a viewpoint that seeks to explain how and why particular photographs were made, what audience they were intended to serve, and, crucially but often forgotten, how they reached that audience, then a different story emerges.[22]

This alternative story captures the technological, social and economic factors that led to the dissemination of modernism through mass photography and also explains the current condition of the hyper-dissemination of architectural imagery today.

Interestingly, however, Ewall's alternative story is one in which a kind of 'critical third way' emerges between the polarization between the independence of art and the economic determinism of a society of consumption, abstraction and figuration, architecture and photography and architects and society. It is an approach to photography in which a reappraisal and revision of photography's heuristic and humanistic nature occurs which allows us it to be a great tool for discovering the real and profound meanings of the constructed world. Object-oriented architectural photography has had fatal consequences for built architecture – the popularizing of the iconic building form being the most obvious example. In contrast to this, a more unrestricted, phenomenological and social concern eager to express and capture the real experience of architecture must, and has, emerged. It is an approach that, at its best, may influence architecture in the ways previous forms of photography have – only this time facilitating a more human focus in design. Supported by artistic sensibility, through the appreciative gaze – and in line with the arguments of John Szarkowski – this emerging approach to architectural photography is less preoccupied with the finished building as an object and more interested in the human and technical processes which precede and produce it.[23]

Notes

1 Robert Elwall, *Building with Light: The International History of Architectural Photography* (London, New York: Merrell Publishers, RIBA, 2004), 8.
2 "The building that can be shown completely in one picture is not worth bothering about." Ezra Stoller, "Photography and the Language of Architecture," *Perspecta* 8 (1963): 43.
3 See Beatriz Colomina, *Architectureproduction* (New York: Princeton Architectural Press, 1988), 9.
4 Alexandra Lange, "Who Pays for the Picture?" *Harvard Design Magazine* 38 (2014): 155.

5 Alberto Campo Baeza, "Borraría muchas cosas de Madrid," *El Mundo*, May 3, 2015, http://www.elmundo.es/madrid/2015/05/03/55466843ca4741722f8b4571.html.
6 See Marshall McLuhan and Quentin Fiore, *The Medium is the Massage: An Inventory of Effects* (New York: Bantam Books, 1967).
7 Joan Fontcuberta, "Hacen falta nacimientos y muertes, también en la fotografía," *El Cultural*, February 17, 2015, http://www.elcultural.com/noticias/arte/Joan-Fontcuberta-Hacen-falta-nacimientos-y-muertes-tambien-en-la-fotografia/7415.
8 See William J. Thomas Mitchell, *What Do Pictures Want?* (Chicago: University of Chicago Press, 2005).
9 Zaha Hadid, "The Photography of Lucien Hervé," in *Lucien Hervé: The Soul of an Architect*, gallery exhibition (London: Michael Hoppen Photography, 199, March 20–May 2). The text by Hadid is displayed on a single full page of the catalogue, not numbered.
10 See Christian Norberg-Schulz, *Intentions in Architecture* (Cambridge, MA: MIT Press, 1965).
11 Michael Leonard, "Humanizing Space," *Progressive Architecture* (1969, April), 128.
12 See Alona Pardo and Elias Redstone, *Constructing Worlds: Photography and Architecture in the Modern Age* (London: Barbican Art Gallery & Swedish Centre for Architecture and Design, 2014).
13 See Pedro Gadanho, "Coming of Age: On the Furtive Shifting Nature of Architectural Photography," in *Shooting Space: Architecture in Contemporary Photography*, ed. Elias Redstone (London: Phaidon, 2014), 231.
14 Gerald Woods, ed., *Eric de Maré, from Art without Boundaries* (London: Thames and Hudson, 1972).
15 Stephen Shore, "Photography and Architecture," in *Sze Tsung Leong: History Images* (Göttingen: Steidl, 2006), 142.
16 John Berger, "Apariencias," in *Otra manera de contar*, ed. John Berger and Jean Mohr (Murcia: Mestizo, 1997), 96.
17 John Szarkowski, "Photographing Architecture," *Art in America* 47 (1959): 86.
18 Robert Campbell, quoted in Akiko Busch, *The Photography of Architecture: Twelve Views* (New York: Van Nostrand Reinhold, 1987), 89.
19 See Hans Danuser, Köbi Gantenbein and Philip Ursprung, *Seeing Zumthor: Images by Hans Danuser* (Zürich: Scheidegger & Spiess, 2009).
20 Hélène Binet, in Peter *Zumthor, Peter Zumthor, Works: Buildings and Projects, 1979–1997* (Baden, Switzerland: Lars Müller Publishers, 1998), 253.
21 "What photography is: is it a mirror, reflecting a portrait of the artist who made it, or a window, through which one might better know the world?" John Szarkowski, *Mirrors and Windows: American Photography since 1960* (New York: Little Brown and Co., 1978).
22 Elwall, *Building with Light*, 8.
23 Szarkowski, "Photographing Architecture," 89.

Bibliography

Berger, John. "Apariencias." In *Otra manera de contar*, edited by John Berger and Jean Mohr. Murcia: Mestizo, 1997, 81–129.

Campbell, Robert. Quoted in *The Photography of Architecture: Twelve Views*, edited by Akiko Busch, 89. New York: Van Nostrand Reinhold, 1987.

Campo Baeza, Alberto. "Borraría muchas cosas de Madrid." *El Mundo*, May 3, 2015, http://www.elmundo.es/madrid/2015/05/03/55466843ca4741722f8b4571.html

Colomina, Beatriz. *Architectureproduction*. New York: Princeton Architectural Press, 1988.

Danuser, Hans, Köbi Gantenbein, and Philip Ursprung. *Seeing Zumthor: Images by Hans Danuser*. Zürich: Scheidegger & Spiess, 2009.

Elwall, Robert. *Building with Light: The International History of Architectural Photography*. London, New York: Merrell Publishers, RIBA, 2004.

Fontcuberta, Joan. "Hacen falta nacimientos y muertes, también en la fotografía." *El Cultural*, February 17, 2015, http://www.elcultural.com/noticias/arte/Joan-Fontcuberta-Hacen-falta-nacimientos-y-muertes-tambien-en-la-fotografia/7415

Gadanho, Pedro. "Coming of Age: On the Furtive Shifting Nature of Architectural Photography." In *Shooting Space: Architecture in Contemporary Photography*, edited by Elias Redstone, 228–231. London: Phaidon, 2014, 231.

Hadid, Zaha. "The Photography of Lucien Hervé." In *Lucien Hervé: The Soul of an Architect*. London: Michael Hoppen Photography, 1998.

Lange, Alexandra. "Who Pays for the Picture?" *Harvard Design Magazine* 38 (2014): 155.

Leonard, Michael. "Humanizing Space." *Progressive Architecture* (1969).

McLuhan, Marshall, and Quentin Fiore. *The Medium Is the Massage: An Inventory of Effects*. New York: Bantam Books, 1967.

Norberg-Schulz, Christian. *Intentions in Architecture*. Cambridge, MA: MIT Press, 1965.

Pardo, Alona, and Elias Redstone. *Constructing Worlds: Photography and Architecture in the Modern Age*. London: Barbican Art Gallery & Swedish Centre for Architecture and Design, 2014.

Shore, Stephen. "Photography and Architecture." In *Sze Tsung Leong: History Images*. Göttingen: Steidl, 2006.

Stoller, Ezra. "Photography and the Language of Architecture." *Perspecta* 8 (1963): 43–44.

Szarkowski, John. *Mirrors and Windows: American Photography since 1960*. New York: Little Brown and Co., 1978.

Szarkowski, John. "Photographing Architecture." *Art in America* 47 (1959): 84–89.

Thomas Mitchell, William J. *What Do Pictures Want?* Chicago: University of Chicago Press, 2005.

Woods, Gerald, ed. *Eric de Maré, from Art without Boundaries*. London: Thames and Hudson, 1972.

Zumthor, Peter. *Peter Zumthor, Works: Buildings and Projects 1979–1997*. Baden, Switzerland: Lars Müller Publishers, 1998.

Part III
Film

7 Absorbing cinematic modernism

From the Villa Savoye to the Villa Arpel

François Penz

It is hypothesized in this chapter that Pierre Chenal's film *Architectures d'Aujourd'hui* (1931) and in particular the scene of Le Corbusier's La Villa Savoye, together with Jacques Tati's La Villa Arpel in *Mon Oncle* (1958), are pivotal to gaining an understanding of a history of architectural modernism through cinema. Chenal's film sequence of the Villa Savoye is a careful translation of Le Corbusier's vision, while in *Mon Oncle* (Tati, 1958), the full force of Tati's humorous satire against modernism is most keenly felt in the way he designed La Villa Arpel.

But let's for a moment fast-forward to the 2014 Venice Biennale where Rem Koolhaas challenged the national pavilions into a common theme, "Absorbing Modernity 1914–2014",[1] and where every nation produced studies of how their infrastructure has been informed by and transformed by modernity. The French Pavilion,[2] curated by architectural historian Jean-Louis Cohen, was entitled *La modernité, promesse ou menace?* (Modernity, Promise or Menace?) and was centrally organised around *Mon Oncle*'s La Villa Arpel (see Fig. 7.2). A short distance away from the French Pavilion was the British Pavilion with an exhibition entitled *A Clockwork Jerusalem* referring to both *A Clockwork Orange* (Kubrick, 1971) and William Blake's poem.[3]

In other words, it would appear as if film has now become central to architect's discourse on modernity.[4] Might it be that in 'Absorbing Modernity', cinema has become a central part of modernity? This would seem paradoxical as cinema has often used modernism as a site of dystopia as argued by Thom Andersen:[5]

> *The most celebrated episode in Hollywood's war against modern architecture is* L.A. Confidential. *Richard Neutra's Lovell house, the first great manifestation of the International Style in southern California, plays the home of Pierce Patchett, pornographer, pimp, prince of the shadow city where whatever you desire is for sale [. . .] Well, the architecture critic of the Los Angeles Times took it seriously. He cited* L.A. Confidential *as some kind of proof that the utopian aspirations of modernist architecture were bogus. He wrote, 'The house's slick, meticulous forms seem the perfect frame for that kind of power . . . Neutra's glass walls open up to expose the dark side of our lives – they suggest the erotic, the broken, the psychologically impure.' So now we know.*

Andersen posits here that cinema is critical of modern architecture because of its association with characters of dubious reputations as in the case of Pierce Patchett and the Lovell House. In other words, he is implying that there is a reflexive relationship between the setting and the action, between the architecture and the film's narrative.

Andersen's remark may at first appear to suggest that this is a one-way relationship, which goes from the action and the characters towards the architecture. The Lovell House found itself tarred with the same brush as the criminals that may occupy it. Actually, the director of *L.A. Confidential* (1997) is an admirer of the Lovell House. 'Curtis Hanson greatly admires the Lovell house. He even gives it a special credit at the end of the film, according it the honorific title favoured by its original owner: the Lovell Health House.'[6] We can only assume that this is 'an accident'. But not so claims the architecture critic of the *Los Angeles Times* that sees the reflexive relationship also going the other way: the Lovell House is the perfect match for a pimp because 'the house's slick, meticulous forms seem the perfect frame for that kind of power'. The implication here is that cinema purposely uses modernism as sites of dystopia and that the two are a perfect match, meaning modernism equals dystopia, at least in cinematic terms, as if Le Corbusier's concept of *La Machine à Habiter*[7] had turned into a vice den. This bold assertion needs closer examination in order to grasp the mechanisms and the forces which led to this perception of the modern movement.

If Andersen provides us here with a working hypothesis, it is Ferro's statement that affords us with the means by which we can start to answer those questions.

> *In its relation to society and history, film was for a long time treated only as a work of art [. . .] Grasping film in its relation to history requires more than just better chronicles of the works or a description of how the various genres evolved. It must look at the historical function of film, at its relationship with the societies that produce and consume it, at the social processes involved in the making of the works, at cinema as a source of history.*[8]

If cinema is indeed an agent, a product and a source of history, then we can argue that part of the history of modernism will have been recorded on celluloid across the twentieth century. It is therefore possible to trace its evolution through films in an attempt to elicit a parallel history – one that is possibly at odds with the history of architecture books on the subject. With the knowledge that cinema can indeed inform us about the history of modernism, we are now in a position to tackle some of the issues raised by Andersen's statement by adopting a largely historical approach, with the Villa Savoye and the Villa Arpel at its heart. We should mention that at the origin of cinema, the camera was trained on everyday spaces as with the Lumière brothers capturing the world around them, or in the filming of ordinary spaces, usually reconstructed in studio in the Biograph Company tradition. They appear on the screen as mere background for the action. We would have to wait until after the First World War to see examples of modernist architecture appear in the movies, such as in the German expressionist movement. The 1920s saw an extraordinary range of experimentation and a true convergence of the arts. Mallet-Stevens became an important figure, making the link between architecture and the cinema, both as a practitioner and a set designer.[9]

But the interest here resides with architects experimenting with cinema and 'real modern' architecture as opposed to film sets. The argument and motivation for filming architecture is provided by architectural historian Sigfried Giedion, who declared, 'Still photography does not capture them clearly. One would have to accompany the eye as it moves: Only film can make the new architecture intelligible.'[10] A number of films stand out as representative of the period: *Die Neue Wohnung* (Hans Richter, 1929), *Architectures d'Aujourd'hui* (Pierre Chenal, 1930) and *Les Mystères du Château du Dé* (Man

Ray, 1928). While we can wonder at Man Ray's poetic surrealist essay and Richter's polished propaganda documentary for a new architecture, it is arguable that *Architectures d'Aujourd'hui*, a collaboration between Chenal and Le Corbusier, responds best to Giedion's claim. It is a film which opens new horizons and breaks new ground, to the point that to this day it resonates as a pioneering work. In that sense, it was a true 'agent' of change, as it paved the way towards new forms of architectural communication.

It is best to take a sequence to illustrate the point. The scene of the Villa Savoye is not only about filming a key icon of the modern movement, but it is the very start of filming modernism as it is the translation of Le Corbusier's Five Points, elaborated in 1927,[11] onto a cinema screen. It is an illustrated reading of his text where the camera underlines the Five Points one by one. It is the transformation of Le Corbusier's writings, mediated by the screen language and for which he had to invent a new visual language. Particularly interesting because it is not a literal translation, as, for example, the order is changed – the original order of the Five Points was pilotis, garden roof, free plan, ribbon window and free facade. So we are invited to a new interpretation of the original text. For example, in the scene of the Villa Savoye, we move from the outside to the inside part of the sitting room by vertical transposition and by visual analogy.

Strikingly, we note how Le Corbusier and Pierre Chenal invented a new film language. In the first part of the sequence, we are in a logic of montage 'à la Eisenstein': there is no human presence; the camera in the free plan is panning around as if looking for a human presence. In a reverse pan, the camera finds the woman who appears at the bottom of the ramp (see Fig. 7.1). At that point, the language changes from montage to continuity editing – from Eisenstein to Griffiths – from visual analogy to a world where we maintain the illusion of continuous action, although time and space has been contracted from real time and real space. Within the same sequence, we experience the transition between Deleuze's concepts of the time-image and the image-movement. In the image-movement, time is measured by bodily movement, while in the time-image, the image and the cuts are not motivated by movements; they are purely optical transitions as opposed to haptical.

Figure 7.1 Villa Savoye – *la promenade architecturale* scene in *Architectures d'Aujourd'hui*. Pierre Chenal, 1930

And, finally, it is in the ramp sequence, *la promenade architecturale*, that we can best appreciate what Le Corbusier was trying to achieve with the film medium. But in order to understand the promenade, we need to briefly evoke Auguste Choisy's *Histoire de l'Architecture* (1899) as he was the first to attempt to retrace the aesthetic motivation of the apparent disorder in the placement of buildings on the Acropolis and to link it precisely to the variable point of view of a mobile spectator. In effect, he explained, elicited and demonstrated the concept of the architectural promenade, as Le Corbusier would later call it. Interestingly, Choisy's lesson was not lost on Eisenstein,[12] as he stated that the 'The Acropolis of Athens is the perfect example of one of the most ancient films'[13] – in other words, for Eisenstein the successive views, or frames, encountered by the mobile spectator in the ascent to the Acropolis would be central to the formulation of his montage theory. Le Corbusier and Eisenstein met in Moscow in 1928 but history doesn't tell if they talked about the Acropolis. But Le Corbusier was certainly influenced by Eisenstein's films, and he claimed that 'Architecture and the Cinema are the only two arts of our time. In my own work I seem to think as Eisenstein does in his films'.[14] Their encounter must have been a major event for Le Corbusier who did not hesitate in his notes to rechristen the Eisenstein film *The General Line* (1928) the straight line (*la ligne droite*), no doubt a reference to his own interest in the regulating lines.[15]

The interpretation offered here is that their encounter will subsequently greatly influence Le Corbusier in his filming of the ramp scene of the Villa Savoye. Indeed, there are a number of inconsistencies in this scene, especially if we consider that Chenal was already an accomplished filmmaker who was very knowledgeable in the continuity editing tradition. Most strikingly, there is no use of the 'shot reverse shot', which would have been considered normal practice in a logic of continuity editing (See Fig. 7.1). As a result, the end of the ramp remains hidden. Furthermore, across shots 2 and 3, the camera crosses the 180-degree line, which is one of the key rules of continuity editing.[16] From shot 3 onwards, we are in a logic of always going forward in the same direction, from right to left. It is suggested here that Le Corbusier was keen to experiment with a new language, which would somehow translate some of his own concerns, in particular, his strong attachment to *la ligne droite* (the straight line), evoked earlier. Therefore, the 'transgressions' encountered in the ramp scene were, for Le Corbusier, necessary devices to chart new territories. Aside from shot 2, the woman always moves in the same direction from the right of the screen towards the left – in effect in a straight line. There is no doubt in my mind that it constitutes both a homage to Eisenstein and a reference to the regulating lines section in *Vers une Architecture*. Overall, the Villa Savoye sequence is the first conscious passage of the language of space, modernism, to the language of the screen through the creation of its own grammar. And as such it is of great historical importance.[17]

In the event, Chenal and Le Corbusier were not to pursue their collaboration. However, modern architecture kept appearing in films, with amongst many others, *A Nous la Liberté* (Clair, France, 1930), *Just Imagine* (Butler, United States, 1930), *Things to Come* (Cameron Menzies, United Kingdom, 1936) and *New Moscow* (Medvedkine, USSR, 1937). Staying within the French context, we must first turn to the wider picture in order to be able to understand the forces which led to the rise of modernism, in particular during post WWII. In France, 'urban planning was an attribute of state power and centralization as well as an instrument of social engineering' remarks Wakeman before adding that it was influenced by 'the extraordinary authority of the Congrès internationaux d'architecture moderne (CIAM) and its promotion of modern aesthetic in the reconstruction years. High modernist utopia pointed the way to the

future'. As a result, '*le peuple* were uprooted and displaced to the outer arrondisse-
ments and suburbs, where brutal demolition of modern housing estates completely
transformed social and spatial arrangements.'[18]

One filmmaker who, in the 1950s, reacted against the demolition and the uprooting
of population was Jacques Tati, in particular with his film *Mon Oncle* (1958). The
story revolves round the clash of two worlds, the old, where Hulot lives and the new
epitomized by the Arpels, Hulot's brother-in-law and sister. In this film, Tati makes
explicit his growing suspicion of modern architecture. If the 1950s can be seen as the
triumph of modernism, Tati certainly expressed strong reservations regarding the type
of built environment developing everywhere at the time. The full force of his humorous
satire is most felt in the way he designed *La Villa Arpel* as it was called. Built in the
Victorine's studio in Nice, Tati explains his take as follows:

> *We had all sorts of architectural reviews and journals which we had gathered. We
> also had some scissors and glue. So I did a montage. I cut some features, a round
> window here, a ridiculous looking pergola there, some garden with a tortuous
> path to give the impression to be bigger than it really was etc., in effect it's an
> architectural "pot-pourri".*[19]

Essentially, we can read La Villa Arpel as the 'anti-Villa Savoye' as every single one of
the Five Points had been either turned on its head or denied (see Fig 7.2).

According to Le Corbusier, the *pilotis* frees up the space underneath the house, *le
jardin passe sous la maison*; this was clearly not the case of La Villa Arpel with a single
column at the entrance. And neither do we have a *toit-jardin*, a garden roof, with the
capacity to double the space of the garden, *le terrain est doublé par l'établissement des
toitures-jardins*. One of the most memorable quotes from the film is Mme Arpel's
assertion to all new visitors that *tout communique* (everything communicates), which
could be construed as Tati's translation for *le plan libre*, the free plan, the absolute
freedom afforded by the plan (*la liberté absolue du plan*). Alas, this too was denied by
a sharp-edged staircase awkwardly placed in the middle of the sitting room and looking 'more like a machine of war for defending one's house against enemies than a
means of communication and access offered to friends'.[20] La Villa Arpel departs even
more radically from the fourth point, *la fenêtre en longueur*, the ribbon window, that

Figure 7.2 Villa Arpel – *Mon Oncle*. Jacques Tati, 1958

has been replaced downstairs by its antithesis, la *porte-fenêtre*, that would have no doubt been applauded by Auguste Perret, while the upstairs is graced with two *fenêtres en rondeur,* round porthole windows. Finally, the lack of *pilotis* to free up the façade results in the denial of the fifth point, *la façade libre*, the free façade, turning it instead into an anthropomorphic façade. The machine for living had been humanized by the round porthole windows acting at night as the eyes of the villa.

Most strikingly, *la promenade architecturale*, the regulating lines, *la ligne droite* (the straight line) had been re-incarnated in the twisted garden path to which we are first introduced through the Arpel family's dachshund, Dacky, in a plaid over-coat, strolling happily along its curved shape – surely the ultimate affront to Le Corbusier! While there is no record of Tati being aware of Le Corbusier's writings or purposely criticizing his oeuvre, he would have no doubt known about his work through his friend and production designer Jacques Lagrange, who designed La Villa Arpel as a pastiche of the International Style, as remarked by 'Le Corbusier' s Maison Jeanneret of 1923, or others by Robert Mallet-Stevens backing on to the same cul-de-sac, which are classic instances of what Lagrange turned into the "Arpel Style"'.[21]

Another remarkable interpretation of the *promenade architecturale* is found in the old part in St Maur where Hulot resides. He ascends the stairs towards his attic home, and while doing so, every step of the way we see part of his body being framed by a disparate range of windows and openings – clearly a metaphor for a human scale architecture. But unlike in the ramp scene of Chenal's Villa Savoye with eight camera set-ups, it is one static, long take in a long shot.

In effect, it is turning on its head Le Corbusier's statement on Arab architecture being best appreciated on foot and where 'you have to walk through a building with a changing viewpoint, to see the articulation of the building deployed' (Le Corbusier 1929, 24). We are not here sharing the point of view of the promenade; instead, we, as spectators, have a clear articulation of a building being deployed in front of our eyes, in all its intricacies, levels and openings. It is an object lesson in a practiced and lived *promenade architecturale* – a new interpretation of Le Corbusier's concept, whereby each successive framing of Hulot's body by a window constitutes a form of cinematic montage edited by the camera.

Figure 7.3 La promenade architecturale: Villa Arpel (left) – St Maur (right) – *Mon Oncle.* Jacques Tati, 1958

This echoes Michel Gondry's remark regarding Tati's filmic inventiveness:

> *What's astonishing with Tati, is that there is a sort of inversion between form and function. For example with Mon Oncle [. . .] what's modern is in fact how he films it, like an extra-terrestrial, like somebody who doesn't accept the screen language conventions [. . .] In other words, the message is not so much in what he tries to say but how he says it. And while his rejection of the modern world may appear somewhat naïve, the means of expressing it are themselves very modern and inventive.*[22]

Indeed, it is particularly true of this scene where Tati creates something novel but out of vernacular architecture – something cinematically modern out of a traditional building – making us notice and value our everyday environment, however banal. It is also a very flat composition of the type favoured by Wes Anderson, in use, for example, in the cinematography of *The Grand Budapest Hotel* (2014), who acknowledges Tati's originality and technique. 'As a filmmaker, he always manages to choose images that are both humorous and aesthetically superb [. . .] it's visually incredibly architecturally structured.'[23]

Tati postulated that *La vie moderne est faite pour les premiers de la classe, ce sont tous les autres que j'aimerais defendre*[24] (Modern life is for those at the top of the class, it's all the others I am trying to defend).[25] Whether Tati's concern with the advent of modernity was naïve or not, he vented it throughout his films, starting with *Jour de Fête* (1949), where François (Tati), the local postman, amidst timeless rural France, attempts to mock the technological advancements in mail distribution in the USA. In *Mon Oncle*, Tati's concern with new technologies is most evident in the kitchen and in the garage scenes. The garage scene in particular is interesting, as it presents a prime example of a film being an agent and product of history (after Ferro) within a cultural and geographical context. In this scene, we see the little dog, having unwittingly activated the 'electric eye', lock the Arpels in their own garage. They call for help and the maid, Georgette, is scared to cross the 'electric beam', a very new technology for the time. But beyond the humorous quality of the situation, we can only understand this scene if we look at the wider context. Georgette, in all likeliness, is from the countryside, where most domestic help would have come from at the time. Indeed, the years surrounding the time *Mon Oncle* was shot saw a surge in rural migration towards large cities, Paris in particular. Pursuing this line of reasoning, we can therefore assume that Georgette would have come from a small village in the countryside where electricity would have been unlikely to appear until after the end of WWII. The advent of electricity in rural France is the subject of a key scene in *Farrebique* (Rouquier 1946), where a family and their old neighbour weigh the pros and cons of such commodity before deciding to adopt it.

Essentially, both scenes are linked by cinema's capacity to contain and deliver aspects of social and cultural history, in this case pertaining to the advent of electricity and modernity in post-war France, and was aptly summed up by Ross as follows:

> *Modern social relations are of course always mediated by objects; but in the case of the French, this mediation seemed to have increased exponentially, abruptly, and over a very brief period of time. If I return throughout the book to the films of Jacques Tati, it is because they make palpable a daily life that increasingly appeared to unfold in a space where objects tended to dictate to people their gestures and*

movements – gestures that had not yet congealed into any degree of rote familiarity, and that for the most part had to be learned from watching American films.[26]

But Tati's *Mon Oncle* breaks away from Rouquier's brand of poetic realism by presenting a 'clownish resistance against modern utopia.'[27] Tati's humour affords a side look at societal issues. One such offering is the theatrical device taking us from the old city where Hulot resides into the new modern world of the Arpels. The transition is always negotiated by a symbolically broken old wall, opening on newly erected housing estates in the background. Those were the 'HLM (Habitations à loyer modéré)[28] apartment blocks rising up at Créteil at Pierre Sudreau's behest.'[29] This sideways look anticipates a future, which was tackled head on by Maurice Pialat in *L'Amour Existe* (1960).

If Tati produced a gentle and humorous commentary on modernism, Pialat's portrayal of the fast growing banlieues was deeply critical. In a reverse move to the New Wave directors who were very much filming within the Paris intra-muros, Pialat's first film concentrated on the suburbs where he grew up. 'For a long time I lived in the suburbs.[30] My first memory is of the suburb'. It was hailed as a masterpiece at the time[31] and according to Fontanel still resonates to this day: 'No conference, meeting or debate on the urban phenomena in 1960s France, can take place today, in France or abroad, without a screening of this film'.[32] It is one of those rare films which transcends boundaries, being part documentary, part sociological and political pamphlet, mixed with fictional elements; the overall impression is both poetic and nostalgic. But the elegant black-and-white cinematography of the modernist housing estate is tainted by Pialat's acerbic voice-over.[33]

> *Now is the advent of the civilian barracks. Concentration camps alike, payable in installments. Urbanism conceived in terms of transport. Poor quality materials deteriotating even before the building is occupied [. . .] The landscape is generally unrewarding, therefore the windows have been suppressed since there is nothing to see.*[34]

L'Amour Existe was released just ten years after *La Vie Commence Demain* (Nicole Vèdres, 1950), itself a very positive outlook on the modern world,[35] including a spirited guided tour of L'Unité d'Habitation in Marseille led by Le Corbusier. But in the intervening years, Le Corbusier's notion of a 'machine for living' had, in Pialat's hands, turned into machines for obliterating! Besides a 'frontal attack' on the HLM, the suburban pavilions (pavillons de banlieue[36]) were not spared either: 'The suburban pavilion can be conceived as a marginal expression of the French's lack of hospitality and generosity. Threatened, it will disappear'.[37]

Pialat was right about the fate of the suburban pavilions; they started to disappear in the 1960s to make way for a much denser urban development just on the outskirts of Paris.[38] This is poignantly documented in the opening sequence of *Le Chat* (Pierre Granier-Deferre, 1971), which consists of a long traveling shot of abandoned and half-demolished pavilions making way for the new high rises. There are films, such as *L'Amour Existe*, which criticize, but others document – perhaps unwittingly – the massive transformation of Paris. *Le Chat* is such as film. Being a fiction film, it uses the physical demolition of the old fabric of the suburb as a metaphor for the disintegration of a couple's life, interpreted by Simone Signoret and Jean Gabin. The torn landscape is a perfect match for their life as a couple, a prime example of spatially organized narrative. But compared to Pialat, Granier-Deferre does not pass judgement on the city, he merely uses it for his own purpose. However, in the process, he records the extraordinary transformation of Cour-bevoie,[39] a small town on the Western side of Paris, which was about to make way for the new *Quartier de La Défense*.[40] The narrative flow goes from the city to the fiction, and there is no return to the city, as it has disappeared. It is a cinema that notes and docu-ments. *Le Chat* has become a *lieux de mémoire*, a site of memory.[41]

English films of the same period provide an equally intriguing interpretation of mod-ernism in films, albeit a different one, partly because the lessons from Le Corbusier appeared to have had, paradoxically, more resonance across the channel, as noted by Colin St John Wilson, 'Corbu – we had allowed him to do our thinking for us.'[42] Indeed, the model of *L'Unité d'Habitation* in Marseille found favours in London as Kite explains, 'In the United Kingdom the celebrated near-realization of such a *ville radieuse* would [have been] the Alton West Housing project (1955–59), Roehampton Lane, London, on its sylvan site sloping towards Richmond Park, built under the LCC team [. . .]'. This proved too good an opportunity to miss for François Truffaut when he was looking to translate Ray Bradbury's dystopian vision in *Fahrenheit 451* (1966). And the Alton West Housing project, England's vision for *la ville radieuse*, became the prime location for the film. Alton West, photographed by Nicolas Roeg, becomes, amongst other locations, the setting of a repressive, totalitarian regime which employs firemen to seek out and burn all books. The implication was that modernism was perceived as authoritarian and denying individuality, and therefore the perfect backdrop for such a fiction.

With Alton West and many lookalike estates, England had generated its own brand of modernism – brutalism – which in turn excited the imagination of other filmmakers, such as Stanley Kubrick in particular with *A Clockwork Orange* (1971). Kubrick's adaptation of the 1962 dystopian novel by Anthony Burgess interpreted the city through its most recent brutalist architecture as a site for the film. In the memorable opening scene, a tramp is brutally killed by Alex and his 'droogs'. Shot in Wandsworth, South London, the underpass is a new kind of modernist space associated with the emerging brutalist architecture of the South Bank; the bevelled edges, the textured concrete, the long perspective would have been a novel visual experience for any visi-tor to the South Bank at the time. But in Kubrick's hands, it acquires a 'double moder-nity', as the space is transformed by the impossibly long shadows and the harsh bluish lighting. The purity of the form has become dehumanized space and the perfect setting for a 'bit of ultra-violence'. And in a curious way, it is as if the original question posed by Reyner Banham in the title of his seminal book *The New Brutalism: Ethic or Aes-thetic?* (1966) was being reignited by Kubrick: the controversy caused by *A Clockwork Orange* resulted in Kubrick's decision to withdraw the film from any UK public screen-ing for nearly thirty years.

If Tati's *Mon Oncle* amounts to a critical appraisal of its time, *A Clockwork Orange* drives a stake through modernism's heart – the sort of blow from which Thamesmead Estate may never have recovered. Yet both films were favourably revisited at the 2014 Biennale – especially Tati's film, which was central to the French Pavilion (see Fig 7.4), both physically and metaphorically. A number of factors are at play here: Undoubtedly Tati's reputation had steadily increased since André Bazin first gave Tati intellectual credentials by writing about *Les Vacances de Monsieur Hulot* (1953) in *Les Cahiers du Cinéma* (1953). François Truffaut singled him out as one of only nine real 'auteurs' in the French cinema of the 1950s.[43] Several books,[44] articles, conferences and film retrospectives have followed in the ensuing decades. It culminated in 2009 with a

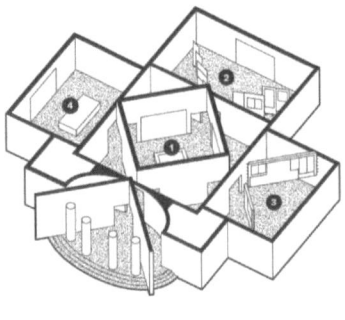

1 Jacques Tati and the Villa Arpel: object of desire or machine of ridicule?
2 Jean Prouvé: constructive imagination or utopia?
3 Heavy prefabrication: economies of scale or monotony?
4 The large housing estate: heterotopia of salvation, or place of reclusion?

Figure 7.4 The French Pavilion at the Venice Biennale 2014. François Penz

major exhibition[45] at the Cinémathèque Française in Paris, where in the words of its then director, Serge Toubiana:

> *Tati has filmed something essential in the course of the 20th century: he filmed the countryside, the everyday life in the countryside (Jour de fête), then he filmed 'la vie pavillonnaire' (Mon Oncle) [. . .] he especially filmed and captured in an ultra- sensitive manner, not unlike a virtuoso seismograph, the passage from the countryside to the city, this epic migration of man and objects from an ancient world towards the modern world [. . .] Everything was changed, the gestures, the trajectories, the atmospheres, the way people dressed. And of course the architecture. The Villa Arpel in Mon Oncle was replaced by the ultramodern buildings of Playtime [. . .] words only can't convey such scale of transformation. That's why Tati's films are mute. They just exist. They are visual noises. They observe with a very precise look, they drill an entomologist gaze onto the human world.[46]*

In some ways, Tati has been elevated to the status of a French national hero, and his films have undergone a process of museification.[47] They can be construed as part of the French identity or *Lieux de mémoire*, alongside 'La Marseillaise, le Panthéon et le 14 Juillet'.[48] Why else would *Mon Oncle* be 'representing France' at the 2014 Venice Biennale, inaugurated by the then French *Ministre de la Culture*, Aurélie Filipetti? The interpretation offered here is that film can be construed as a form of practice of architecture involving mechanisms of production as well as consumption and reception, then and now. While there is much to gather and understand from the 'data culled' by filmmakers, we can also reflect on Ferro's concept of film as an 'agent' which, within this context can be interpreted along the lines of Keiller's assertion, 'In films, one can explore the spaces of the past, in order to better anticipate the spaces of the future.'[49] This implies that a film's reception may have an active participation in present and future events. The reading put forward in this chapter converges here with the official message in the preface[50] to the exhibition catalogue, which states, 'the French Pavilion expresses the promises of architecture to enable the construction of the world of tomorrow.'[51]

And so *Mon Oncle's* Villa Arpel has become part of the representation of the future, a future, or even, an acceptable future. Looking back from when Tati and Lagrange conceived the Villa Arpel in 1957, it has gradually developed into an active agent, 'an agent of change'. Overtime this characteristic has gathered more and more momentum with each exhibition;[52] its aura and appeal widening with time. Admittedly, it is visually a rather stunning model, especially the garden design with its multicolored gravel, a direct reference to the celebrated surrealist landscaping design by Gabriel Guévrékian in 1926 for the Villa Noailles designed by Mallet-Stevens.[53] And let's face it, if it wasn't aesthetically pleasing, it would never have made it to the Venice Biennale in 2014.

Still it is paradoxical that Jean-Louis Cohen, the curator of the French Pavilion, chose not to reference the Villa Savoye,[54] given that he is one of the most preeminent scholars of Le Corbusier. Instead, Cohen stresses the historical importance and significance of the Villa Arpel. 'Both in its interior layout and exterior form – and extending through to the garden with its relentless geometry – this fictitious house echoes the projects that modern architects were still unable to carry out in France at the time', while adding that, 'the Villa Arpel, more than 50 years later, continues to symbolize the dream of a life made easier by the machine, but which sometimes turns to farce'. So unlike the critic of the *Los Angeles Times* (commenting on *L.A. Confidential*), Cohen argues that the utopian aspirations of modernist architecture portrayed in films – that also includes Godard[55] – have to be taken seriously.

It is an acknowledgement that cinema significantly contributed to the absorption of modernity over the last one hundred years or so and that film as a key technology of sight has induced major changes in our perception of architecture. And it is useful to conclude with a quote from Antonioni:

> *We know that underneath an image shown, there is another image, which is more faithful to reality, and that underneath that second image there is still another one, and then one more. Right up to that true image of that absolute and mysterious reality that nobody will ever see.*[56]

In other words, when we look at the image of the Villa Arpel in the Venice Biennale (see Fig. 7.5), it evokes a multitude of narrative layering and conjures up other images,

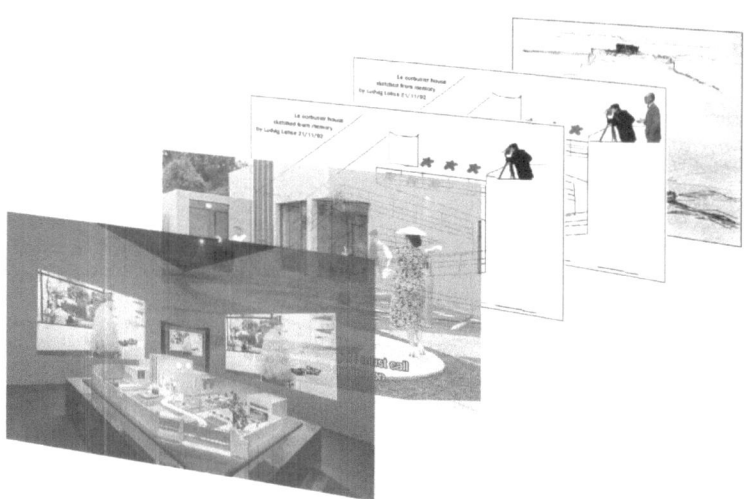

Figure 7.5 The narrative layering from 2014 to the Acropolis. François Penz

amongst which the Villa Savoye, Pierre Chenal and Le Corbusier, with perhaps in the distance 'for the eyes that can see', the Acropolis, Eisenstein and Choisy!

Notes

1 Rem Koolhaas, *Fundamentals: 14th International Architecture Exhibition. La Biennale Di Venezia 2014.* (Venice: Marsilio, 2014).
2 Ibid., 64.
3 Ibid., 68.
4 This argument departs from the traditional discourse on modernity and early cinema as expressed, for example, by Tom Gunning 'Cinema both as a practice and as a force [. . .] played a central role in the culture of modernity . . . cinema metaphorised modernity' Tom Gunning, "Modernity and Cinema," in *Cinema and Modernity*, ed. Murray Pomerance (New Brunswick, NJ: Rutgers University Press, 2006, p. 302) – the author here concentrates instead on the relationship between film and the modern movement, specifically architectural modernism.
5 Voice-over in Thom Andersen's *Los Angeles Plays Itself* (US, 2003).
6 Ibid.
7 Le Corbusier, *Vers une Architecture* (Paris: Editions Flammarion, 2008).
8 Marc Ferro, "Film as an Agent, Product and Source of History," *Journal of Contemporary History* 18, no. 3 (1983): 358.
9 Odile Vaillant, "Robert Mallet-Stevens, Architecture, Cinema and Poetics," in *Cinema and Architecture Méliès, Mallet-Stevens, Multimedia*, ed. François Penz and Maureen Thomas (London: British Film Institute, 1997, 28–33).
10 Sigfried Giedion, *Construire en France, construire en fer, construire en béton* (Paris: La Villette, 2000), 92.
11 Werner Oechslin and Wilfried Wang, "Les Cinq Points d'une Architecture Nouvelle," *Assemblage* 4 (Oct. 1987): 86.
12 Russia was of the first country where a translation of Auguste Choisy's, *Histoire de l'Architecture* (Paris: Gauthier-Villars, 1899) was undertaken. The first volume of the translation appeared in 1906, the second in 1907 and this book had some impact on the formation of the Russian constructivist architects of which Eisenstein was one.
13 Sergei Eisenstein, "Montage and Architecture," *Assemblage* 10 (1989): 117.
14 This is the inscription that Le Corbusier wrote when he offered *L'Art décoratif d'aujourd'hui* (1925) to Eisenstein in 1928. Jean-Louis Cohen, *Le Corbusier and the Mystique of the USSR, Theories and Projects for Moscow, 1928–1936* (Princeton: Princeton University Press, 1992), 72.
15 Ibid.
16 See the author's earlier articles on the subject for more information: François Penz, "Notes and Observations Regarding Pierre Chenal and Le Corbusier's Collaboration on Architectures d'Aujourd'hui (1930–31)," in *Design and Cinema: Form Follows Film*, ed. B. Uluoğlu, A. Enşici and A. Vatansever (Newcastle upon Tyne: Cambridge Scholars Press, 2006), 149–167; François Penz, "L'ombre de l'Acropole – La Villa Savoye construite par le cinema," in *L'invention d'un architecte. Le voyage en Orient de Le Corbusier*, ed. Roberta Amirante, Burcu Kütükçoglu, Panayotis Tournikiotis and Yannis Tsiomis (Paris: La Fondation Le Corbusier, 2013), 407–413.
17 The author has written extensively on the subject. For more information, see Ibid.
18 Rosemary Wakeman, *The Heroic City: Paris, 1945–1958* (Chicago: University of Chicago Press, 2009), 17.
19 In an article published by the *Journal des Monuments Historiques*, Marie-Anne Sichère, "Jacques Tati: Où-est l'architecture?" in *Monuments Historiques: 'La dernière séance'* 137 (Février-Mars 1985), 86–87.
20 Alphonse Karr, *300 Pages: Melanges Philosophiques* (Whitefish, MT: Kessinger Publishing, 1861), 198–199.
21 David Bellos, *Jacques Tati: His Life and Art* (London: Harvill, 1999), 207.
22 Author's translation from the French. See: Macha Makeieff and Stéphane Goudet, *Jacques Tati: Deux Temps, Trois Mouvements* (Paris: Editions Naïve, 2009), 243.
23 Ibid.

24 He further elaborated that 'Je ne me moque pas du modernisme. J'essaye simplement de faire en sorte que la nouvelle société soit plus conviviale et que l'on ne devienne pas de simples machines [. . .] Mais il faut faire attention à ne pas nuire à la convivialité, notamment par l'architecture. Si les gens peuvent communiquer, c'est parfait, mais il semble que nos sociétés développent la confusion et ne communiquent plus du tout'. Ibid.

25 Philippe Mary, "Le cinéma de Jacques Tati et la « politique des auteurs »," *Actes de la recherche en sciences sociales* 1, no. 161–162 (2006): 42.

26 Kristin Ross, *Fast Cars / Clean Bodies: Decolonization and the Reordering of French Culture* (Cambridge, MA: MIT Press, 1995), 11.

27 Wakeman, *The Heroic City*, 239.

28 The French for social housing (HLM is short for Habitat à Loyer Modéré).

29 Wakeman, *The Heroic City*.

30 'Longtemps j'ai habité la banlieue. Mon premier souvenir est un souvenir de banlieue.' The first line of the voice-over is reminiscent of Proust's first sentence in *À la recherche du temps perdu* 'Longtemps, je me suis couché de bonne heure'.

31 In 1961, it received both the 'Prix Louis Lumière' and the 'Lion de Saint-Marc, Venice Film Festival'.

32 Rémi Fontanel, "Le Centre et la marge: cadrer l'extérieur. La représentation filmique de la banlieue dans L'Amour existe de Maurice Pialat," in *La Ville au cinéma*, ed Julie Barillet, Françoise Heitz, Patrick Louguet and Patrick Vienne (Artois: Artois Presses Université, 2005), 208.

33 This sequence was shot in les Courtillières (Pantin, North East of Paris) a suburb built between 1954 and 1964 by Emile Aillaud (architecte) and Rieti Fabio (colorist), the same conceptual team responsible for Chanteloup-les-Vignes where *La Haine* (Kassovitz, 1995) takes place.

34 Author translation from the French: Voici venu le temps des casernes civiles. Univers concentrationnaire payable à tempérament. Urbanisme pensé en termes de voirie. Matériaux pauvres dégradés avant la fin des travaux [. . .] Le paysage étant généralement ingrat, on va jusqu'à supprimer les fenêtres puisqu'il n'y a rien à voir.

35 Aside from Le Corbusier, it included interviews from André Gide, Frédéric Joliot-Curie, Irène Joliot-Curie, Daniel Lagache, Pablo Picasso, Jean Rostand and Jean-Paul Sartre.

36 In between the two world wars, there is a massive extension of small, individual pavillons, often built on illegal land, just outside the Paris intra-muros.

37 'Le pavillon de banlieue peut être une expression mineure du manque d'hospitalité et de générosité du Français. Menacé il disparaîtra'.

38 A number of other films should be referenced here for fulfilling similar function: *La Crise du Logement* (Jean Dewer, 1956), which tackles the issue of the slum clearance; *Le Joli Mai* (Chris Marker, 1963) for its last part when families are being interviewed and enthusiastically move into the new HLM leaving behind crowded slums; and *La Ville Bidon* (Jacques Baratier, 1973), which takes a very political and satirical view of the advent of the new HLM developments.

39 Pierre Granier-Deferre shot *Le Chat* in Courbevoie (10km West of Paris) and by some strange coincidence, Pialat in *L'amour existe* says 'for a long time I have lived in this area of Courbevoie' [Longuement j'ai habité ce quartier de Courbevoie].

40 Tjebbe van Tijen wrote extensively on this film in an article entitled *Le chat: cinematic psycho-geography of Paris Courbevoie/La Défense*. On the urban transformation of Paris in the 1970s, he noted, 'The new super métro the RER was then constructed with its devastating effects on the urban structure: the razing of suburban areas like Courbevoie for the new high rise business centre La Défense and at the other end of the line the digging of what Parisians called at that time 'le grand trou' (the big hole) what is now Le Châtelet/Les Halles'. Reference: http://limpingmessenger.wordpress.com/2010/09/28/le-chat-cinematic-psycho-geography-of-paris-courbevoiela-defense/ [September 28, 2010].

41 Pierre Nora, "Between Memory and History: Les Lieux de Memoire," *Representations* 26 (Spring 1989): 7–24.

42 Stephen Kite, "Softs and Hards: Colin St. John Wilson and the Contested Visions of 1950s London," in *Neo-Avant-Garde and Postmodern: Postwar Architecture in Britain and Beyond*, ed. Mark Crinson and Claire Zimmerman (New Haven: Yale University Press, 2010), 75.

43 Bellos, *Jacques Tati*, 202.

44 Difficult to estimate precisely the number of books currently available on Tati and his films – Cambridge University Library holds around twenty but there are many more – but they span at least sixty years, from Geneviève Agel, *Hulot parmi nous* (Paris: Les Éditions du Cerf, 1955) – to Jeanne Vronskaya, *Jacques Tati* (Frankfurt am Main: Literarischer Europäer, 2015).

45 A full scale model of the Villa Arpel was recreated at le CENTQUATRE-PARIS in the nineteenth arrondissement – in partnership with the Cinémathêque exhibition. Prior to that, in 2003, as part of the exhibition *Tatirama* at the Netherland Architecture Institute in Rotterdam – part of the Rotterdam Film Festival – a one-tenth scale model of the Villa Arpel was also exhibited.

46 Author's own translation from the French of a quote from Serge Toubiana. See Makeieff and Goudet, *Jacques Tati*.

47 The interaction between cinema and museum was articulated in architectural terms by Mumford as early as 1937. For him, the cinema intervenes in museographic culture by conserving what cannot otherwise be kept in existence – the reproduction of life on film acts as a perspective on what otherwise would be lost. In particular, cinema preserves and conserves precious aspects of everyday life – the very core of what makes up a culture.

48 Nora, *Between Memory and History*, 1989.

49 Patrick Keiller, "The City of the Future," *City: Analysis of Urban Trends, Culture, Theory, Policy, Action* 7, no. 3 (2003): 384.

50 The preface was co-authored between Laurent Fabius, the then foreign office minister, and Aurélie Filipetti, the culture minister (Institut Français 2014).

51 Ibid.

52 Tatirama in 2003 in Rotterdam and Milan, CENTQUATRE-PARIS in 2009 and Venice Biennale in 2014 – although there might have been others.

53 Makeieff and Goudet, *Jacques Tati*, 106.

54 There is no mention of Le Corbusier in the exhibition, which was organized as follows: 'Running throughout the entire pavilion, a film composed of archival images and clips from films such as those by Jacques Tati or Jean-Luc Godard provides a dynamic visual narrative that ties the four sequences of the exhibition together: Jacques Tati's Villa Arpel / Panels from Jean Prouvé's curtain walls / Heavy prefabricated concrete panels by the engineer Raymond Camus / The Cité de la Muette in Drancy, by Euge`ne Beaudoin and Marcel Lods' (Institut Français 2014). However, the exhibition was accompanied by a book presenting 101 buildings by French architects from 1914 to 2014, where Tim Benton wrote a short piece on La Villa Savoye (Jean-Louis Cohen, Vanessa Grossman and Nikola Jankovic, *Modernity, Promise or Menace? France: 101 Buildings 1914–2014* [Paris: Éditions Dominique Carré, 2014], 54).

55 As part of the exhibition, a twenty-minute film, projected simultaneously in each gallery, *Modernity, Promise or Menace?* (Teri Wehn Damisch 2014) included sequences from *2 ou 3 choses que je sais d'elle* (Jean-Luc Godard 1966) – as well as *Mon Oncle*.

56 Michelangelo Antonioni, Carlo Di Carlo, Giorgio Tinazzi and Marga Cottino-Jones, *The Architecture of Vision: Writings & Interviews on Cinema* (New York: Marsilio Publishers, 1996), 63.

Bibliography

Agel, Geneviève. *Hulot parmi nous*. Paris: Les Éditions du Cerf, 1955.

Antonioni, Michelangelo, Carlo Di Carlo, Giorgio Tinazzi and Marga Cottino-Jones. *The Architecture of Vision: Writings & Interviews on Cinema*. New York: Marsilio Publishers, 1996.

Bazin, André. *Mr Hulot et le temps*. Paris: Cahiers du Cinéma, 1953.

Bellos, David. *Jacques Tati: His Life and Art*. London: Harvill, 1999.

Choisy, Auguste. *Histoire de l'Architecture*. Paris: Gauthier-Villars, 1899.

Cohen, Jean-Louis. *Le Corbusier and the Mystique of the USSR, Theories and Projects for Moscow, 1928–1936*. Princeton: Princeton University Press, 1992.

Cohen, Jean-Louis, Vanessa Grossman, and Nikola Jankovic. *Modernity, Promise or Menace? France: 101 Buildings, 1914–2014*. Paris: Éditions Dominique Carré, 2014.

Eisenstein, Sergei. "Montage and Architecture." *Assemblage* 10 (1989): 110–131.

Ferro, Marc. "Film as an Agent, Product and Source of History." *Journal of Contemporary History* 18, no. 3 (1983): 357–364.

Fontanel, Rémi. "Le Centre et la marge: cadrer l'extérieur. La représentation filmique de la banlieue dans *L'Amour existe* de Maurice Pialat." In *La Ville au cinéma*, edited by Julie Barillet, Françoise Heitz, Patrick Louguet and Patrick Vienne, 208. Artois: Artois Presses Université, 2005.

Giedion, Sigfried. *Construire en France, construire en fer, construire en béton*. Paris: La Villette, 2000.

Gunning, Tom. "Modernity and Cinema." In *Cinema and Modernity*, edited by Murray Pomerance, 302. New Brunswick, NJ: Rutgers University Press, 2006.

Institut Français – Press Kit 2014 (consulted on 24 March 2016): http://www.institutfrancais. com/sites/default/files/press_kit_international_exhibition.pdf

Karr, Alphonse. *300 Pages: Melanges Philosophiques*, 198–199. Whitefish, MT: Kessinger Publishing, 1861.

Keiller, Patrick. "The City of the Future." *City: Analysis of Urban Trends, Culture, Theory, Policy, Action* 7, no. 3 (2003), 384.

Kite, Stephen. "Softs and Hards: Colin St. John Wilson and the Contested Visions of 1950s London." In *Neo-Avant-Garde and Postmodern: Postwar Architecture in Britain and Beyond*, edited by Mark Crinson and Claire Zimmerman, 75. New Haven: Yale University Press, 2010.

Koolhaas, Rem. *Fundamentals: 14th International Architecture Exhibition. La Biennale Di Venezia 2014*. Venice: Marsilio, 2014.

Le Corbusier. *Les oeuvres complètes*, Vol 1. Zurich: Erlenback-Zurich, 1929.

Le Corbusier. *Vers une Architecture*. Paris: Editions Flammarion, 2008.

Makeieff, Macha and Stéphane Goudet. *Jacques Tati: Deux Temps, Trois Movement*. Paris: Editions Naïve, 2009.

Mary, Philippe. "Le cinéma de Jacques Tati et la « politique des auteurs »." *Actes de la recherche en sciences sociales* 1, no. 161–162 (2006): 42–65.

Nora, Pierre. "Between Memory and History: Les Lieux de Memoire." *Representations* 26 (Spring 1989): 7–24.

Oechslin, Werner and Wilfried Wang. "Les Cinq Points d'une Architecture Nouvelle." *Assemblage* 4 (Oct. 1987): 82–93.

Penz, François. "Notes and Observations Regarding Pierre Chenal and Le Corbusier's Collaboration on Architectures d'Aujourd'hui (1930–31)." In *Design and Cinema: Form Follows Film*, edited by Belkis Uluoğlu, Ayahn Enşici and Ali Vatansever, 149–167. Newcastle upon Tyne: Cambridge Scholars Press, 2006.

Penz, François. "L'ombre de l'Acropole – La Villa Savoye construite par le cinema." In *L'invention d'un architecte. Le voyage en Orient de Le Corbusier*, edited by. Roberta Amirante, Burcu Kütükçoglu, Panayotis Tournikiotis and Yannis Tsiomis, 407–413. Paris: La Fondation Le Corbusier, 2013.

Ross, Kristin. *Fast Cars / Clean Bodies: Decolonization and the Reordering of French Culture*. Cambridge, MA: MIT Press, 1995.

Sichère, Marie-Anne. "Jacques Tati: Où-est l'architecture?" In *Monuments Historiques*: 'La dernière séance' 137 (Fèvrier-Mars 1985): 86–87. Paris: Ministère de la Culture, Éditions de la Caisse Nationale des Monuments Historiques et des Sites.

Vaillant, Odile. "Robert Mallet-Stevens, Architecture, Cinema and Poetics." In *Cinema and Architecture: Méliès, Mallet-Stevens, Multimedia*, edited by François Penz and Maureen Thomas, 28–33. London: British Film Institute, 1997.

Vronskaya, Jeanne. *Jacques Tati*. Frankfurt am Main: Literarischer Europäer, 2015.

Wakeman, Rosemary. *The Heroic City: Paris, 1945–1958*. Chicago: University of Chicago Press, 2009.

8 Fragmented fluidity

A possible future for spatial theory and praxis in filmic form

Graham Cairns

The 1980s saw the emergence of a form of architecture premised on the conceptual frameworks of deconstruction. Aesthetically, it borrowed from the constructivist architects of the early twentieth century and spatio-temporally it drew parallels with the discontinuity cinema epitomized by Sergei Eisenstein. The lead proponent of this early cinematic parallel with late twentieth-century architecture was Bernard Tschumi. In reaching back to the discontinuity cinema of the 1920s as a conceptual framework with which to scaffold contemporary architectural theory and practice, Tschumi was retreading terrain continually worked throughout the twentieth century by architects of different hues. In placing the relationship between film and architecture back on the agenda, however, his choice of an early twentieth-century filmmaker to do so was revealing. The architecture of the time, it seemed, lacked a contemporaneous reference point of the cinema.

Within ten short years of Tschumi's identification of filmic influence on architecture, however, the conceptual climate of avant-garde architecture had changed. Derrida's influence, and with it the fragmentary aesthetic of the deconstructivist architects, was replaced as the conceptual aesthetic *de rigueur*. What replaced it was a Deleuzian ensemble of ideas and terms more easily applied to the folding and fluid forms being created by the end of the 1990s by the architectural parametricists. Although parametricism, still at work on the architectural psyche but less emotive in its self-promotion today, never fully sought to bolster itself through cinematic parallels, there were occasional references made by people such as Greg Lynn and Lars Spuybroek in those 'early years'. Just as the deconstructivists lacked a contemporary filmic movement to conceptually draw upon, so too did the parametricists. What this chapters suggests is that just at the historical point of interchange between these two competing architectural approaches, a formal tendency was emerging in film that could, and still can, offer an aesthetic and conceptual frame of reference that both deconstructivists and parametricists could fruitfully call upon.

It is suggested here that films such as *Russian Ark*, *Run Lola Run*, *Zidane: A Twenty-First Century Portrait* and Mike Figgis' *Timecode* potentially offer a vantage point, formal language and conceptual approaches with which we can complement contemporary architectural production, aesthetics and theory. They are films that offer a point of reference for explaining and visualizing the spatial effects both deconstructivists, but more fully, parametricists, sought and seek to produce. They also offer a set of spatial practices that can inform the spatial planning and choreography of architectural space. In addition, they give us a spatio-temporal conceptual system that can enrich and further the theoretical discussions instigated by both the architecture of deconstruction and parametricism. In doing so, these films offer contemporary architecture the opportunity to continue to develop under the influence of a medium it has been enthralled with since its inception over one hundred years ago.

The filmic context

Directed by Alexander Sokurov, *Russian Ark*, 2002, is a single-shot, ninety-six-minute film recorded on uncompressed high-definition video by one continually moving camera.[1] Coordinating over two hundred actors to move, interact and weave narratives together through a single unified time and space sequence, it completely rejects standard narrative techniques and the goal-oriented rules of continuity. It layers spaces, events and movements in a fluid and sophisticated filmic construction. Characters, narratives and times all merge with one another, flow alongside each other and evolve out of each other.

By contrast, the Tom Twyker film *Run Lola Run*, 1996, is narratively conventional and acutely goal based. However, the way in which this film evolves towards the resolution of that goal is anything but conventional by overlaying three similar event sequences in three distinct spatio-temporal strata.[2] Structurally, then, it opts for a literal tripartite layering of narratives, spaces and, apparently, simultaneous times. In each of its three sequences, these techniques facilitate both a unifying, and a breaking, of space and time that can remind us of discontinuity's principal techniques. These ruptures, however, are secondary to the film's circular, repetitive and mutating spatio-temporal structure that folds nuances into our reading of the sequential stories as they emerge.

Directed and produced by Douglas Gordon and Philippe Parreno in 2006, *Zidane: A Twenty-First Century Portrait* follows one man through the continuous and unbroken storyline of a single game of football. Rejecting standard narrative and plot line in favour of a "real-time experience", it takes the enclosed space of the football pitch and fragments it through multiple extreme close-ups. Those fragments are never fully disjointed, however, with the proximity of the footballer's body imposing a unity of space throughout. It is yet another recent film that employs advanced technologies to create a spatio-temporal filmic experience that is multi-layered, intricately fragmented and complexly continuous.[3]

In these films, it is possible to discern a number of analogous approaches to filmic space and time – approaches in which multiple streams of simultaneous events, movements and spaces overlap, co-exist, interchange and morph. Although it may not be possible to categorize them as part of a coherent school, they do exhibit a spatio-temporal filmic repertoire that has much in common with Derridian-like concepts of fragmentation and Deleuzian ideas of space and time. To consider this repertoire in more detail, Mike Figgis' *Timecode*, 2000, is a useful point of departure, as it offers up readings that potentially align it with both the deconstructivist and parametricist tendencies in architecture. It was also a film produced as the precise moment when the architectural tide, or fashion, was shifting towards the parametric.

Used to synchronize and log digitally recorded video and film, SMPTE timecoding is a technology woven into contemporary televisual production. It replaces the manual process of logging start and end times of shots, with shot-logging software in computers or in cameras themselves.[4] The name of this film is then a direct reference to the technicalities of contemporary filmmaking. It is shot on four cameras recording simultaneously without interruption for ninety minutes. Each of these continuous 'films' is presented

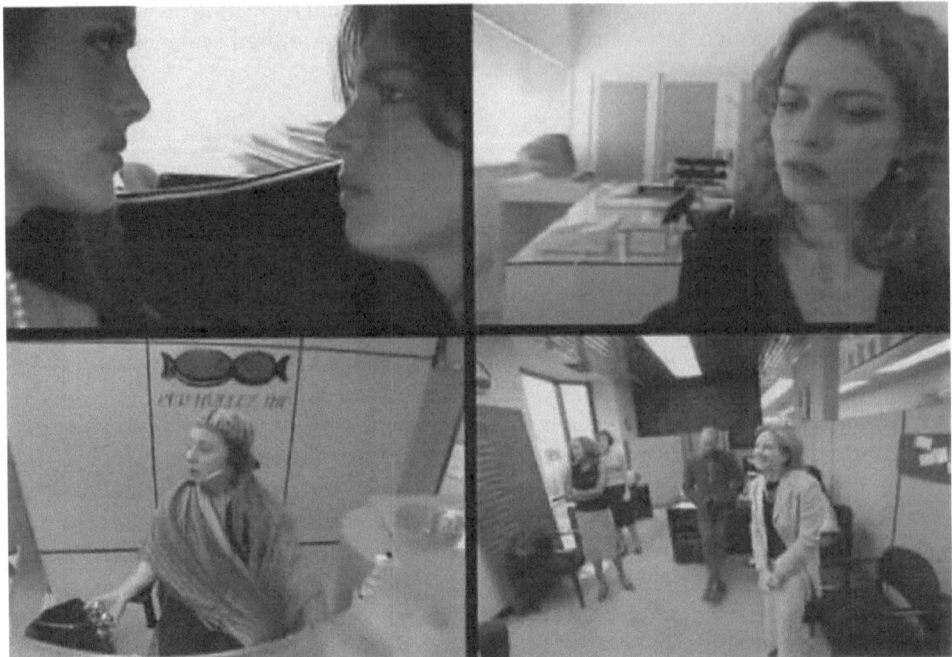

Figure 8.1 Still from Timecode. Mike Figgis, 2000

simultaneously on a screen divided into four quadrants which turns the standard unified filmic aesthetic into something based on juxtaposition and layering (Fig. 8.1).

Timecode, then, is a form of filmic experiment that commercial cinema rarely makes and, for the viewer, it is a complex, multifarious visual experience impossible to follow in conventional ways.[5] The reasoning behind this experimental approach is self-referentially explained in the film itself as "an attempt to rescue resurrect montage as a radical techniques capable of challenging the norms of visual artistic production and consumption".[6] The film, then, made its own formal experimentation a key component of its own analysis.

Derrida, Tschumi and architectural convention

The breaking of convention, the inversion of hierarchy and the criticism of staid, commercial practice at the heart of *Timecode* echoes with the early projects, theories and writings of Bernard Tschumi in multiple ways, most obviously, with his 'piece de resistance', *Parc de la Villette*, Paris.[7] As with *Timecode*, this project was presented as a reworking of architecture's 'most basic conventions and principles'. In particular, the Derridian challenges to 'false hierarchies' and 'prejudices' inherent in the linguistic conventions of structuralism were key and would remain central to Tschumi's architecture from that point onwards.[8] For Tschumi, these linguistic prejudices find their architectural parallel in the discipline's one principal convention: the hierarchical importance given to built architectural form over the free use of spaces by people.[9]

Tied into this hierarchical dominance of building over activity, or space over event, was the tendency for most buildings to only ever facilitate one use over another.[10]

Figure 8.2 Follie at Parc de la Villette. Bernard Tschumi, 1982–1998. Graham Cairns

Developing an approach based on the juxtaposition of events and spaces, however, also led to an aesthetic of juxtaposition, and it was this that Tschumi graphically sought to justify, and theoretically support, through reference to the cinema of Sergei Eisenstein. In projects such as *Parc de la Villettte*, single buildings are composed of multiple different fragments, violently forced together in an architecture of aggressive asymmetry and formal rupture. In its planning, too, *Parc de la Villette* asymmetrically superimposed one set of infrastructures on another. As a result, it represented the most emblematic deconstructive building of the 1980s (Fig. 8.2).

Peter Eisenman and the architecture of defamiliarization

The deconstructed and juxtaposed aesthetic evident in both the architecture of Tschumi and Figgis' *Timecode* was also manifest in the aesthetic developed by Peter Eisenman in architectural projects of this period. As with Tschumi, he collaborated closely with Derrida. His early theoretical texts challenged the entire humanist tradition upon which he argued twentieth-century architecture was based. In particular, he proposed the rejection of a 'universal' thinking associated with neo-Classicism.[11] In this regard, Eisenman drew upon the Derridian rejection of logocentricism and its apparent dominance at the heart of Western philosophical thought.[12] He also echoed Tschumi in his examination and rejection of architectural conventions and could be argued to have foreshadowed Figgis' analogous reaction to cinematic convention in *Timecode*.

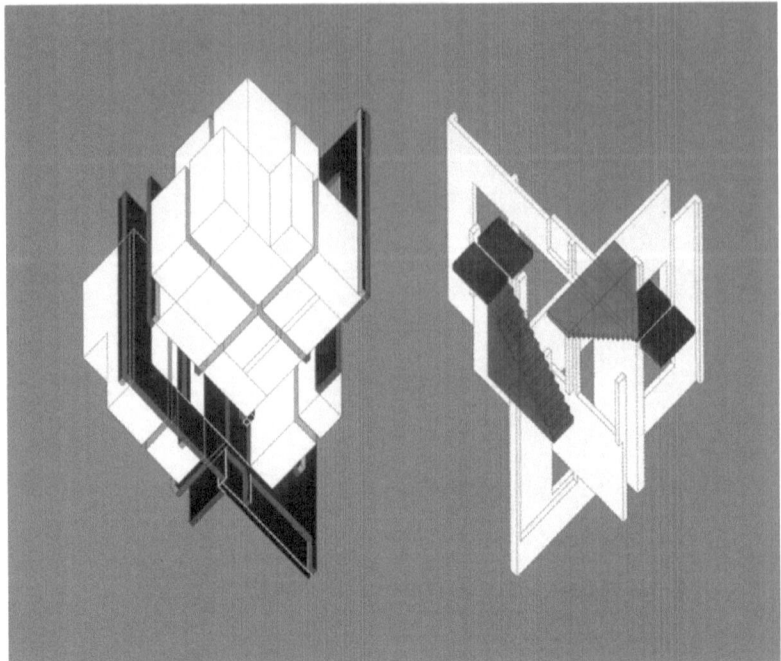

Figure 8.3 House VI. House of Cards, Peter Eisenman

Nowhere is this more evident than in the experimental houses documented in the 1987 publication *House of Cards*.[13] In the Houses I–V, documented in this text, Eisenman claims to take the formal and structural vocabulary of architecture apart. In short, he 'deconstructs' it. Walls are pulled apart from each other, partitions hover above the floor, floors in turn expand and contract beyond the edges of the building and stairs lead to the emptiness above one's head. It is an architecture that remains inhabitable, but which sought to provoke a double reading of architectural vocabulary and use (Fig. 8.3). Krauss describes it in the formalist terms of the European avant-garde, whilst Tafuri resorts to linguistics and defines it as "a semantic game which brings the syntactic system (of architecture) to the foreground".[14]

Eisenman does not shy away from either of these descriptions but places particular emphasis on the propensity of both readings to turn architecture into a self-referential game with its own formal, structural and aesthetic tropes. It is a strategy he sees as leading architecture away from what he defines as the external impositions of rules, laws and conventions into an autonomous, experimental, freely creative discipline. Indeed, he has described this type of architectural investigation as examinations of architecture's "interiority" – an examination of its own discourse.[15] This internal deconstructive discourse clearly echoed Derrida and Tschumi at the level of architectural theory, but also produced the fragmentary, disjunctive aesthetic highlighted in Tschumi´s *Parc de la Villette* and Figgis' *Timecode*. The work each of these authors, then, can be said to gravitate around a self-proclaimed 'rejection of convention' on the one hand and an aesthetic of formal fragmentation on the other.

From formal properties to narrative in *Timecode*

The confrontational and disjunctive aesthetic of *Timecode* is, at its most basic level, produced through its division of the screen into four quadrants. Simultaneously presenting multiple continuous films, Figgis creates an optical experience that is inherently ruptured and even chaotic. This sense of optical fragmentation, however, is supplemented by a number of other characteristics and traits. The cameras used are handheld and, as a result, the images produced are shaky. This random movement means that the composition formed in each quadrant is constantly changing as the camera twists from one angle to another in ways clearly analogous to Tschumi's and Eisenman's formal disjunctures. Furthermore, there tends to be a significant amount of movement by the actors themselves in each quadrant – a characteristic that leads to the creation of a film in which visual complexity and apparent randomness dominate the overall visual effect.

These formal properties not only produce a multiple, fragmentary and overloaded visual effect on screen, they allow Figgis to play with multiple spatial and temporal simultaneities. These layered simultaneities question the authority of the normally dominating sole camera viewpoint and replace it with what has been called a "distributed gaze".[16] It is a multi-focal viewing approach that creates a multi-faceted image and soundscape – a landscape that is impossible to absorb in all its detail in a single viewing. Instigating a viewing experience that involves the obligatory 'selective attention' of the viewer, it reminds us of Maurice Merleau-Ponty's arguments in the *Phenomenology of Perception*.[17] Aylish Wood describes it in slightly different terms. Faced with a multiplicity of narrative pieces, spatial fragments and temporal simultaneities when watching *Timecode*, the cinematic viewer develops "distributed attention".[18]

With this distributed attention, viewers are theoretically free to concentrate on whatever fragment of the multi-sensorial image and soundscape they choose. The viewer who tries to follow the related events is obliged to jump from one space and event to another in a continually altering and alternating 'act of reading'. The narrative reading initiated by Figgis thus becomes potentially random, multiple and never ending as one image is compared against another. It is a cinematic structure that potentially sets up what may be referred to in Derrida's deconstructive terms as *a continual deferment of* meaning.[19] In standard post-structuralist terms, the narrative 'act of reading' becomes open ended, and we are presented with Roland Barthes' notion of "the death of the author" and in Umberto Eco's "role of the reader".[20]

In the final analysis of *Timecode's* structure, however, this characteristic is not taken to the inevitably chaotic end it suggests. Figgis subtly 'cues' viewers into following a preferred, if not totally controlled, narrative sequence; he fades and synchronizes dialogue and sound effects throughout the film. Aural activity in one quadrant is manipulated to subtly impose itself at key moments, whether that is through an explicit increase in volume or through directing particular actors to avoid conversation at specific moments. Similarly, the image-scapes are controlled by emphasizing or reducing the action taking place in the various quadrants, which is a directorial effect that is sometimes reinforced by a blurring of focus. Despite this reinsertion of author control, however, the visual effect remains fragmentary and overloaded, and the narrative reading remains open to subversion. The reader remains free to roam across the screen constructing his or her own narrative readings.

Form and narrative in the work of Tschumi

In introducing the potentially open-ended narrative reading of *Timecode*, we find yet another analogy evident with the work of Tschumi and his approach to architecture as "event-space", most notably explored in *The Manhattan Transcripts*.[21] (Tschumi, 1981) Attempts to describe the *experience* of architecture from the perspective of events, the 'architectural projects' of *The Transcripts* employed a graphic system of photographs, diagrams and plans. Laid out as independent frames running vertically and horizontally, it compositionally echoes *Timecode's* split screen. The basic supposition argued by Tschumi here is that any architectural experience involves spaces, events and movement.[22] Space is represented by the plan, events by the photographs and movement of protagonists by diagrams. The aim is not to design a building, but to explore the nature of how we experience architecture, what becomes important, then, are not built forms but actions.

As with *Timecode*, however, the real importance of the *Transcripts* lies not in its actions, but in how they are presented aesthetically and formally and, by extension, how they are read and experienced. Aesthetically, each frame tends to be characterized by fragmentation and deconstruction. The photographs are snapshots, the diagrams show conflicting lines of movement whilst the drawings are distorted perspectives or even collages. Formally, these individual frames, set out in vertical and horizontal strips, are normally presented in groups of twelve and are intended to be too multifarious to take in at once (Fig. 8.4)

Given both the lack of formal hierarchy in their presentation, and their independence as isolated images, they can be read in both a vertical and a horizontal sequence and, of course, randomly.[23] Not only is the formal layout analogous to the screen in *Timecode* but so too is the author's intention with regard to how they are read. They are visually complex and dynamic individual frames, which, as a whole, lead to visual overload and potentially open-ended narrative readings, which is something also identifiable in Tschumi's most documented built work, *Parc de la Villette*.

Visually, the most defining aspect of *la Villete* are its thirty-five individual red buildings based on a 10.8 m cube.[24] Dotted across the site according to a grid plan, they

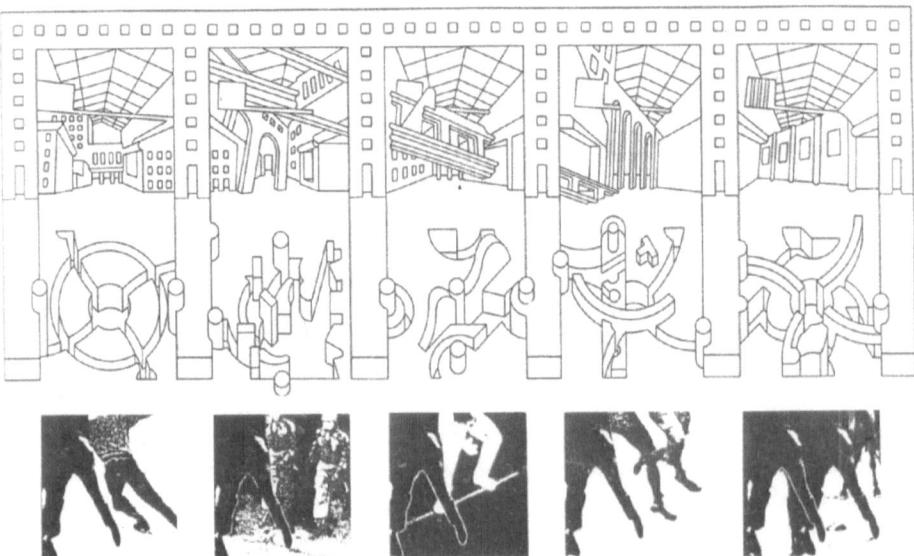

Figure 8.4 Diagrams from Manhattan Transcripts, 1981. Bernard Tschumi

are visible from wherever one stands. With the buildings aligned in this way, one can clearly see the formal characteristics of *The Manhattan Transcripts* reappear and, if one considers *Timecode*, here too certain formal echoes re-emerge. However, the grid plan is not the only, or even the dominant, planning system used by Tschumi. It is only one of three systems that are superimposed one on top of the other. On top of the *grid* is laid the system of *lines*: two lineal axis pedestrian routes running north-south and east west, together with secondary *lines* such as random curvilinear pathways. Further superimposed on top of the *grid* and the *lines* is a system of *surfaces*: different zones such as landscaped gardens, playgrounds and open green spaces (Fig. 8.5). It is a planning strategy that Tschumi refers to in his usual vocabulary as "disjunctive".[25]

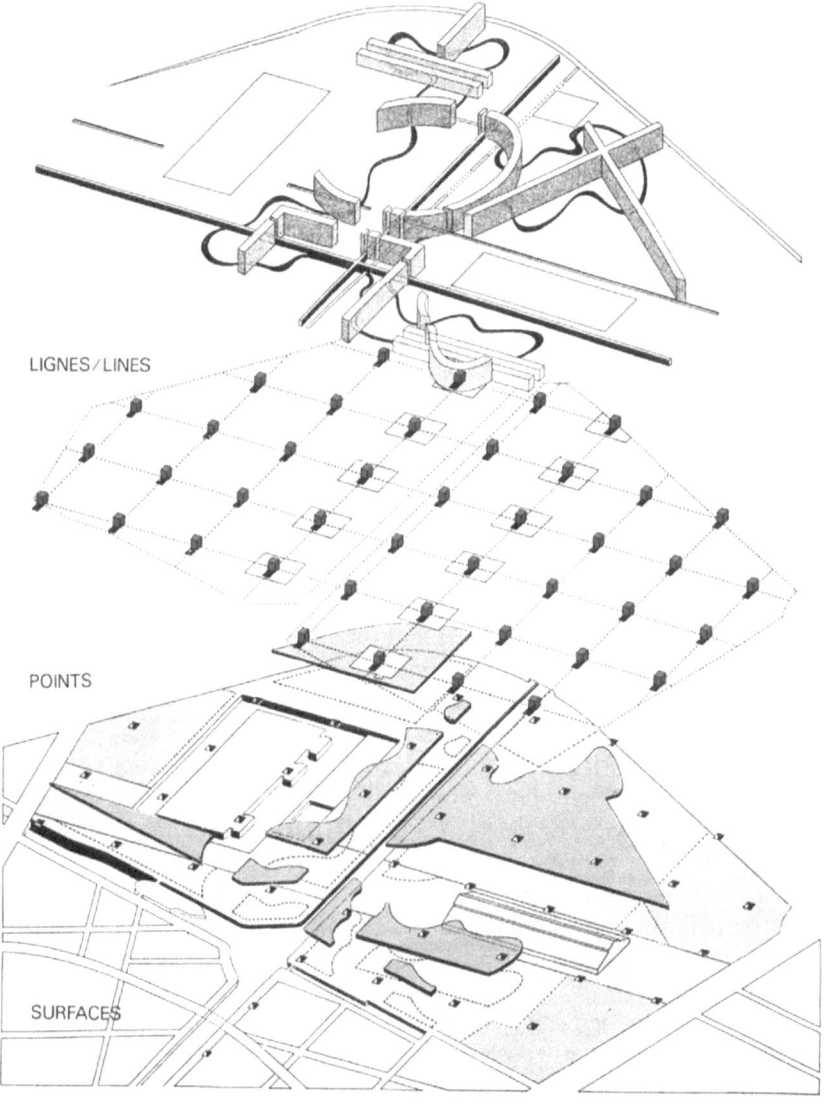

Figure 8.5 Diagrammatic Plan of Parc de la Villette, 1982. Bernard Tschumi

The aesthetic result of this juxtaposition of equally valid spatial layouts was a series of violent ruptures as the walkways of the line system clashed with the follies of the grid system, which, in turn, ran across the system of surfaces. Narratively, too, something similar occurs. Instead of using one of these systems, and thus having a space with clearly defined pathways and use patterns, Tschumi attempts to create a movement pattern in which users can chop and change the sequence of their journey across the park. Here, when we walk along a path we inevitably cross a surface which, in turn, leads us to a folly on the grid. There is no one spatial sequence or narrative that dominates our use, and we are free to move between the parks various spatial trajectories. Echoing Figgis, Barthes and Derrida in architectural terms, Tschumi allows users the freedom to change and move around the park according to their own preferences and decisions. He relinquishes author control.

Timecode and the overlaying of Deleuzian space

The analogies we find between *Timecode*, *The Manhattan Transcripts* and *Parc de la Villette* open up multiple possible relationships between the architecture of deconstruction and contemporary film. There is, however, one particular difference that leads us away from the underlying ideas of Derrida and Tschumi and direct us towards the realms of Deleuze and Eisenman. It is a difference that opens up a whole new realm of development for the cinematic-architecture relationship reignited in the 1980s.

We find in the work of Tschumi and Figgis an interest in the 'inversion of convention', the 'rejection of standard hierarchies', the loosening of author control and an apparent aesthetic of fragmentation. However, we also have in *Timecode* a multi-layered presentation of events, spaces and characters that doesn't wholly rupture their relationship, but rather allows them to blend and merge. In *Parc de la Villette*, Tschumi set out three planning systems, which, although independent, are deliberately intended to clash and produce violent rupture. In the *Manhattan Transcripts*, too, the imposition of events in alien spaces is claimed to provoke aggressive and fragmentary interchange. Tschumi defined this "architecture of disjunction" as a deliberate "act of violence".[26]

Although *Timecode* similarly separates out time, space and events, and thus produces a visual effect definable as chaotic and fragmented, Figgis does not treat these constitutive factors as wholly independent and draws back from allowing them to 'violently' clash. Partly due to his use of 'cueing' techniques, and partly due to the continual simultaneity of their presentation, Figgis unifies spaces and events through layering rather than truly rupturing them through disjunction. He does not fragment so much as multiply and thus allows his simultaneous independent narrative strands to morph and diverge fluidly at different points throughout the film.

Whilst the aesthetic experience for the viewer may be one of distortion and apparent chaos, which can remind us of the cinema of Vertoz and Eisenstein referenced by Tschumi, the cinematic footage of Figgis is, in reality, more pliable. A character may leave one quadrant as the camera shifts, but simultaneously reemerges in another as his or her particular narrative carries on seamlessly and apparently unaware of the camera. In these instances, the division between the split screens is not so much a line, a border or a rupture, but rather a fold in a homogenous smooth cinematic space.

Similarly, a phone call can be made by a character in one quadrant that introduces a character in another into the dominant 'attended' storyline. In these moments, the border again disappears as separate actions and spaces are given equal importance without fragmentation or rupture; they simply co-exist simultaneously across a cinematic space bridged by multiple cameras. Imagery in *Timecode*, then, is not broken, it is smoothed and opened to a process of blending and divergence. Space and time are not fragmented, but rather intertwined into something equally analogous, albeit in formal terms, to the 'smooth' spaces of Deleuze and Guattari, as it is to the ruptured spaces of Tschumi.[27]

In a sense, *Timecode* merges Deleuze's notion of the movement-image and the time-image as laid out in *Cinema 1* and *Cinema 2*.[28] The *movement-image*, inherent in the continuous filming of the events in each quadrant, is blended with the *time-image* as Figgis allows the past, the present and future to interpenetrate through merger: the events in one quadrant that lead to consequences in another being analogous to past events seen through flashback. Similarly, future events that result from the action in one section evolve in another quadrant as the action moves on, in a separate space and time, analogous to the flashforward.

In her interpretation of Deleuze's arguments on film, Claire Colebrook has argued that both the movement-image and the time-image are central to one of his principal arguments about the medium in general: its ability to be disruptive to standard thought. Deleuze's thinking on cinema, she suggests, is not concerned with explanation, but rather an open-ended examination of its 'mechanisms' and, by extension, an interest in pushing those mechanisms to their limits, with the aim being to instigate a reconsideration of how we think and conceive.[29] Thus when we see *Timecode* pushing the technical possibilities of film to its limits, and consequently instigating a reconsideration of the relationship between space and time, we see film operating in the way celebrated by Deleuze: as a medium capable of altering its own discourse and those of other fields.[30]

A Deleuzian reading of *Timecode*, then, is not solely based on an interpretation of its fluid representation of time and space. On the contrary, it penetrates much deeper into the substrata of some of his principal ideas. Nevertheless, in our cinematic-architectural context, it is precisely this spatio-temporal presentation that is of interest. Given the way film progresses and its narratives emerge and evolve without rupture or apparent hierarchy, it may be described as forming a series of Deleuzian 'plateaus'. The film's actions evolve and morph across the multiply split cinema screen, slipping from one realm to another as if passing over smoothed-out boundaries and levels rather than hitting the borders between quadrants.

Taking the application of Deleuze's terminology even further, this reading of *Timecode* reveals spaces, times and events that are unstable but which are not fragmentary; they operate as a series of evolving "differences" which, in their continual emergence and fading, can be described as representative of the notion of "immanence".[31] Furthermore, one could use the notions of "smooth" space, "nomadic" events and even the idea of "swarming" in literal translations which help describe the narrative, visual and formal effects of *Timecode's* filmic techniques and language.[32] It is then, for multiple reasons, that the film is more fruitfully and fully understood through the prism of Deleuze than the fragmentary concepts of Derrida and deconstruction.

The emergence of Deleuzian space in the architecture of Eisenman

Clearly, reading *Timecode* as a morphing of spatio-temporal phenomena rather than as disjunctive fragmentation begins to open up a significant distance between our understanding of this type of film and the spatial effects created by deconstruction.

However, it also pulls us closer to the later architecture of Eisenman who, after drawing heavily on Derrida in the 1980s, began to reference Deleuze and Guattari with ever-more regularity in the 1990s. In the 1990s, Eisneman's 'internal' reconsideration of the architect's thinking and design tools began to move beyond the then recognized boundaries of the discipline itself. Employing the generative logic of computer modelling, he termed his form of parametricism as "the conceptualizing of other methods".[33] Picking up the then emerging technology and thinking of a new generation, the computer, he argued, 'conceptualizes and draws differently to the human mind and hand'. As a result, it produces an architecture that cannot be created by designers working and thinking through traditional tools. The now well-established practices of parametricism were seen as fomenting a new architectural vocabulary – one based on notions of morphing, self-generating randomness and what we may call a form of 'immanence'.

In projects such as the *Max Reinhardt Haus*, Berlin, 1992; *Haus Immendorff*, Düsseldorf, 1993; the *BFL Software Headquarters*, Bangalore, 1997; and the *Church for the Year 2000*, Rome, he has exploited the then emerging formal trends of folds, waves, bends, transgressions and transformations. In the *Haus Immendorff* project, for example, the "generative power of the computer" was used to simulate soliton waves which, "due to abrupt changes in depth or subterranean seismic patterns, form non-linear interactions".[34] Imitating the way soliton waves undergo constant change within singular aqueous forms, the algorithm employed in this 'creative process' simulated an architectural form intended to seamlessly blend and randomly morph.[35]

The same formal properties that, as Mario Carpo[36] has most notably suggested, have been emerging for decades are evident in the *Church for the Year 2000* project in which an analogy with liquid crystals is used to explain its folding form.[37] Repeated again in the *BFl Headquaters* project, this liquid crystal metaphor is combined with an interest in the mandala to create an algorithm that results in another folding, overlaid and inter-lapping series of spaces and forms.[38] In a more recent project, the City of Culture in the Galician region of Spain, the computer modelling employed in the design process combined and distorted data which replicated the medieval street plan of Santiago de Compostela – a Cartesian grid and the topography of the surrounding hills (Fig. 8.6).

Figure 8.6 'The City of Culture', 2011, Spain. Peter Eisenman

In these later projects, the fragmentation of deconstruction and its cinematic analogies found in Soviet Montage are replaced by Deleuzian notions of 'immanence', 'nomady' and 'deterritorialization', which, in turn, find visual echoes in a Deleuzian reading of the fluid multi-screen simultaneity of *Timecode*.

The analogies here with *Timecode* are twofold. First, we see the evolution of a fluid, folding aesthetic and sequences and, second, we have a process in which the author has once again 'relinquished' complete control. In Figgis' case, this occurs in the more open-ended readings he is prepared to countenance, but it is also seen in his generative technique. The employment of four simultaneous camera shots sets up an inherently complicated and multi-faceted filmic algorithm of its own, over which the director can never have complete control. Consequently, what we see potentially emerging is a new architecture-filmic analogy in which questions of aesthetic, form and form generation are all reframed by parametricism in Deleuzian terms. It is a reconfiguration of our reading of the formal associations between architecture and film that is potentially of direct relevance to numerous practitioners and theorists today.

Greg Lynn has argued for an architecture of "animate design": an architecture whose design process is fundamentally characterized by the dynamic forces of generative computer processing. Defined as "the co-presence of motion and force at the moment of formal conception", it is central to his methodology and the forms that result.[39] Referencing film in this regard, Lynn suggests that "the cinematic model of space" has tended to separate dynamic and static events and spaces in ways typical of standard architectural thought. Film, he says, "sees motion as a series of frames through a static location". Reflected in the typical employment of computer modeling in design, the rendering of a 'walkthrough' animation that gives a sense of a building's interior, it is a mindset that only allows for movement to be introduced into architecture post-design process. For Lynn, parametricism allows 'movement' to be a component factor of the form creation process itself and, furthermore, facilitates the production of smoothly blended forms for which the most up-to-date cinematic morphing effects serve as a precedent and analogy.[40]

In addition to referencing film, however, Lynn repeats the arguments put forward here in a number of other ways. First, in his search for an architecture of folding and fluid spaces, he has always been explicit. In *Folds, Bodies and Blobs: Collected Essays*, 1998, he directly rejected the fragmentary and disjunctive approach of deconstruction in favour of a more fluid conceptualiztion of architecture. In making the case for this transition, he discusses one of Eisenman's most important projects, *The Wexner Centre for the Arts*, Ohio. Normally categorized as "disjunctive" due to a number of fragmentary characteristics, Lynn suggests its equally prevalent "continuous" characteristics, most notably its roof, should lead to it being reframed as a Deleuzian "model of simultaneity".[41]

This explicit referencing of Deleuze is the second way in which we can interpret Lynn as echoing the arguments laid out here, and it is repeated in *Animate Form*. In this text, Lynn associates an interest in Deleuze to a whole generation of architects "weary of the representational critiques of Deconstruction" and more in tune with a new age of organic forms or "Blobitecture".[42] What Lynn seems to be hinting at here then, is the consensus around a coherent set of ideas and theories, and the formal similarities in output, that are necessary to identify a coherent movement in architecture – a movement it could be suggested that finds a filmic counterpart in the experimental work of directors such as Figgis, Twyker and Sokurov.

Less explicitly Deleuzian, but nonetheless centered on the creation of an architecture of the foldable and the pliable, is Lars Spuybroek. At the forefront of promoting parametricism, Spuybroek returns to Eisenman's interest in the diagram in the formulation of his own generative process in which the diagram becomes a conceptual input/output device that "swallows, restructures and ejects matter".[43] (Spuybroek, 2006) What he claims his computer-generated diagrams produce is, as with Lynn, malleable and fluid architectural forms in which plans, volumes, floors, screens and surfaces blend. It is an architecture that again "rejects conventional mechanistic experience in favour of a more visceral environment in which action, perception and vision are synthesized".[44]

Less associated with Blobitecture but certainly fully aligned with its parametric processes for years now, is Thom Mayne of *Morphosis*. Although coming late to the unpredictable generative capabilities of the contemporary computer-aided design, the work of *Morphosis* adapted it to its full potential quickly. Mayne has defined the work of *Morphosis* as based on "questioning the notion of boundary" and "oscillating between the notion of the inside and outside". It also questions the idea of "centre and periphery" and what he suggests is its "undefinable inverse".[45] In his employment of such terminology, Mayne begins to give a sense of the folding intricacies found in projects such as the University of Cincinnati Recreation Building: a project created using the computer to "manipulate, rescale, stretch, amend, subtract and pry apart single architectural objects" in the creation of a new generation-folded architecture.[46]

Other contemporary architects who share these parametric tendencies, to a greater or lesser extent, include the partners of the now defunct Foreign Office Architects. At their most celebrated project, the Yokohama Port Authority building, their design process involved a computer algorithm fed with multiple data streams in the creation of another multiply folded architectural form. It is an architectural typology that they described as producing "an experience in which whatever direction you walk, you never have the sense of returning to the same spot".[47] In short, they described it as an architectural experience potentially analogous to the viewing experience of *Timecode*, *Run Lola Run* or *Russian Ark*.

Similar traits can be found in the work of MvRdV in projects such as Barendrecht Church, 1993, Sloterpark Swimming Pool, 1994 and the Hasselt Villa, 1996. Not only does MvRdV use complex data sets in the generation of form, but it produces folding and pliable "landscape" buildings that can merge architectural form with its surroundings.[48] It is formally analogous to the terminology of Deleuze and aesthetically analogous to the films mentioned here. It also permits a reading of how we walk through, across and over architecture in ways that recall the viewing of the experimental films of Figgis, Twyker and Sokurov: continuous, emerging and in a state of continual flux. As with the architecture of Spuybroek, Mayne, FoA, Eisenman and the numerous others now working the same furrow, their buildings become parametric generations that echo the fluidity of experience and the potential fluidity of the contemporary strand of film discussed here.

Conclusion

What emerges from all of this is a set of possible parallels between the formal and spatial concepts that underlie parametricism with those that can be found in a number of recent high-profile productions from commercial and non-commercial film sources. Certainly, in both the architectural and filmic spheres, new types of fluid, unpredictable and extremely complex notions of space, time, form and aesthetics have recently emerged and share important similarities. These new ideas can be said to coalesce around themes identifiable in spatial readings of the work of Gilles Deleuze. Consequently, we find ourselves presented with the opportunity to redefine, or at least consider the redefinition, of the relationship between film and contemporary parametric architecture in new terms: in terms of the folding and supple formalism extractable from the Deleuzian paradigm and evident in films such as *Russian Ark*, *Run Lola Run* and *Zidane: A Twentieth-Century Portrait* and the film we have discussed here in detail, *Timecode*. Should the parametricists, now at the height of their acceptance and influence thus far, choose to build a filmic theory for their work, it would not be unexpected and would not represent a radical shift. On the contrary, it would represent a continuation of a filmic-architectural conceptual, aesthetic and formal relationship that has benefited architecture for over a century.

Notes

1 Louis Menashe, "Filming Sokurov's Russian Ark: An Interview with Tilman Büttner". *Cineaste* 28, Issue 3 (2003): 21–23.
2 For an extensive overview of the film, see Ingeborg Majer O'Sickley, "Whatever Lola Wants, Lola Gets (Or Does She?): Time and Desire in Tom Tykwer's Run Lola Run". *Quarterly Review of Film & Video* 19 (2002): 123–131.
3 Manohla Dargis, "Portrait of the Artist as a Global Soccer Star". *New York Times: Culture*. Film Review. October 23, 2008. p. 10.
4 Ashley Wood, "Encounters at the Interface: Distributed Attention and Digital Embodiments". *Quarterly Review of Film and Video* 25 (2008): 219–229.
5 Ibid., 221.
6 This logic is exported in a monologue by one of the films protagonists, Ana Pauls, a deliberately pretentious avant-garde European filmmaker played by Mia Maestro. Whilst making a pitch to Red Mullet executives for the production of her latest film, she lays bare the complex theoretical underpinnings of her proposal, which is clearly *Timecode* itself.
7 Amongst his other realized projects one finds the *Interface Flon Railway and Bus Station*, Lausanne, the *Alfred Lerner Hall Student Centre*, Columbia University, and *Le Frenosy National Studio for the Contemporary Arts* in Tourcoing. For an overview, see Samantha Hardingham and Kester Rattenbury, *Bernard Tschumi: Parc de la Villette: SuperCrit No. 4* (London: Routledge, 2011).
8 Michelle Lamont, "How to Become a Dominant French Philosopher: The Case of Jacques Derrida". *American Journal of Sociology* 93, Issue 3 (1987): 602.
9 Developing his architecture of space and event equivalence, he argued, amongst other things, that there can be there can be "no space without events, nor architecture without program". See Bernard Tschumi, *Event-Cities* (London: MIT Press, 1994).
10 It was in this register that he playfully queried whether we could "deconstruct" this one dimensional view of building and use by, for example. . . . "riding bicycles in launderettes or pole vaulting in lift shafts?" Ibid.
11 John Whiteman, *Investigations in Architecture: Eisenman Studios at GSD: 1983–85* (Cambridge, MA: Harvard University Press, 1986), 7.
12 Jacques Derrida, "Deconstruction and the Other". Interview with Richard Kearney. In: R. Kearny (ed), *Dialogues with Contemporary Continental Thinkers* (Manchester: Manchester University Press, 1984), 123.
13 With texts by Rosalind Krauss, Manfredo Tafuri and Eisenman himself, *House of Cards* documents Eisenman's early works and uncovers the theoretical logic threaded through their formal experimentation with the architectural language of the Modern Movement. Each author underlines the architect's attempts to "wake architecture from its slumber" through a series of projects that defamiliarize our engagement with dwelling and indeed, space itself.

14 Manfredo Tafuri, "Peter Eisenman: Meditations of Icarus". In: Peter Eisenman (ed), *Houses of Cards* (New York: Oxford University Press, 1987), 167.
15 Peter Eisenman, *Diagram Diaries* (London: Thames and Hudson, 1999), 29.
16 Wood, "Encounters at the Interface," 220.
17 Arguing that the perception of any phenomena involves our concentration on an "object of attention", and the concomitant lack of attention on secondary stimuli, Merleau-Ponty identifies that much of the data in our sensorial field gets "backgrounded"; relegated to a secondary level of attention. See Maurice Merleau-Ponty, *The Phenomenology of Perception* (London: Routledge, 1962), 68.
18 Wood, "Encounters at the Interface," 224.
19 William Ray, *Literary Meaning: From Phenomenology to Deconstruction* (Oxford: Basil Blackwell Ltd, 1984), 147.
20 Umberto Eco, *The Role of the Reader* (Indiana: Indiana University Press, 1979), 5.
21 Bernard Tschumi, *Manhattan Transcripts* (London: Academy Editions, 1981), 6.
22 Ibid., 9
23 Ibid., 10.
24 Each cube, referred to as a folly, was broken into a number of conflicting components which could be removed, distorted or extended. They thus take on a compositional appearance of abstract and fragmented cubic sculptures. None of these were given any specific function by the architect, with the intention being to challenge the hierarchy of events and spaces by provoking conflicts between uses, forms, activities and buildings.
25 Bernard Tschumi, "Parc de la Villette, Paris". In: A. Papadakis, C. Cooke and A. Benjamin (eds), *Deconstruction Omnibus* (New York: Rizzoli, 1989), 175–181.
26 Bernard Tschumi, *Architecture and Disjunction* (Cambridge, MA: MIT Press, 1997), 122.
27 Gilles Deleuze and Felix Guattari, *A Thousand Palteaus: Capitalism and Schizophrenia* (London: Continuum, 1988), 478.
28 Gilles Deleuze, *Cinema 2: The Time-Image* (London: The Anthlone Press, 1989), 34.
29 Claire Colebrook, *Gilles Deleuze* (London: Routledge, 2002), 34.
30 Ibid.
31 Andrew Ballantyne, *Deleuze and Guattari for Architects* (London: Routledge, 2007), 50.
32 Ibid., 18.
33 Peter Eisenman, "A Conversation with Peter Eisenman". In: A. Zaera-Polo (ed), *Peter Eisenman* (Madrid: El Croquis, 1997), 13.
34 Ibid., 132.
35 Describing this project, Eisenman argues that walls, floors, ceiling and roof merge in a non-humanly controlled generative process to create a fluid, evolving and blending series of volumes that aesthetically and formally have more in common with notions of smooth space and merging plateaus. See Eisenman, "A Conversation with Peter Eisenman".
36 See Mario Carpo, *The Digital Turn in Architecture, 1992–2012* (Chichester: Wiley, 2013).
37 Gradually distorting as it emerges out of the ground, rises and moves along the floor, *The Church of the Year 2000* is an architecture of multiple layers and overlaps said to reflect the "phased distortions of a liquid crystal in its nematic state".
38 Eisenman, "A Conversation with Peter Eisenman," 163.
39 Greg Lynn, "Animate Form". In: C. Jencks (ed), *Theories and Manifestos of Contemporary Architecture* (London: Wiley-Academy, 2006), 328.
40 Greg Lynn, *Folds, Bodies and Blobs: Collect Essays* (Brussels: La Lettre volée, 1998), 110.
41 Ibid., 116.
42 Greg Lynn, *Animate Form* (New York: Princeton Architectural Press, 2011), 40.
43 Lars Spuybroek, "Machining Architecture". In: C. Jencks (ed), *Theories and Manifestos of Contemporary Architecture* (London: Wiley-Academy, 2006), 351.
44 Peter Zellener, *Hybrid Space: New Forms in Digital Architecture* (New York: Thames and Hudson, 1999), 112.
45 Spuybroek, "Machining Architecture," 351.
46 Thom Mayne, "Morphosis: Connected Isolation". In: Thorn Mayne (ed), *Architectural Monographs*. No. 23 (London: Academy Editions, 1993), 11.
47 Farshid Moussavi and Alejandro Zaera-Polo, "Code Remix 2000". In: C. Jencks (ed), *Theories and Manifestos of Contemporary Architecture* (London: Wiley-Academy, 2006), 338.

48 Richard Levene and Fernando Márquez (eds.), "Stacking and Layering". In: *El Croquis: MvRdV* (Madrd: El Croquis editorial, 1997), 52, 56 and 118.

Bibliography

Ballantyne, Andrew. *Deleuze and Guattari for Architects*. London: Routledge, 2007.

Carpo, Mario. *The Digital Turn in Architecture, 1992–2012*. Chichester: Wiley, 2013.

Colebrook, Claire. *Gilles Deleuze*. London: Routledge, 2002.

Dargis, Manohla. "Portrait of the Artist as a Global Soccer Star". *New York Times: Culture*. Film Review. October 23, 2008. p. 10.

Deleuze, Gilles. *Cinema 2: The Time-Image*. London: The Anthlone Press, 1989.

Deleuze, Gilles and Guattari, F. *A Thousand Palteaus: Capitalism and Schizophrenia*. London: Continuum, 1988.

Derrida, Jacques. "Deconstruction and the Other". Interview with Richard Kearney. In: R. Kearny (ed), *Dialogues with Contemporary Continental Thinkers*. Manchester: Manchester University Press, 1984, 123.

Eco, Umberto. *The Role of the Reader*. Indiana: Indiana University Press, 1979.

Eisenman, Peter. "A Conversation with Peter Eisenman". In: A. Zaera-Polo (ed), *El Croquis: Peter Eisenman*. Madrid: El Croquis Editorial, 1997, 13.

———. *Diagram Diaries*. London: Thames and Hudson, 1999.

Hardingham, Samantha and Rattenbury, Kester. *Bernard Tschumi: Parc de la Villette: SuperCrit No. 4*. London: Routledge, 2011.

Johnson, Philip and Wigley, Mark. *Deconstructivist Architecture*. New York: The Museum of Modern Art, 1988.

Krauss, Rosalind. "Death of a Hermenutic Phantom: Materialization of the Sign in the Work of Peter Eisenman". In: Peter Eisenman (ed), *House of Cards*. Oxford: Oxford University Press, 1987, 168.

Lamont, Michelle. "How to Become a Dominant French Philosopher: The Case of Jacques Derrida". *American Journal of Sociology* 93, Issue 3 (1987): 602.

Levene, Richard and Márquez, Fernando (ed). "Stacking and Layering". *El Croquis: MvRdV*. Madrd: El Croquis Editorial, 1997, pp. 52, 56 and 118.

Lynn, Greg. *Animate Form*. New York: Princeton Architectural Press, 2011.

———. "Animate Form". In: C. Jencks (ed), *Theories and Manifestos of Contemporary Architecture*. London: Wiley-Academy, 2006, p. 328.

———. *Folds, Bodies and Blobs: Collect Essays*. Brussels: La Lettre volée, 1998.

Majer O'Sickley, Ingeborg. "Whatever Lola Wants, Lola Gets (Or Does She?): Time and Desire in Tom Tykwer's Run Lola Run". *Quarterly Review of Film & Video* 19 (2002): 123–131.

Mayne, Thom. "Morphosis: Connected Isolation". In: Thorn Mayne (ed) *Architectural Monographs*. No. 23. London: Academy Editions, 1993, p. 338.

McGilligan, Patrick. *Robert Altman: Jumping Off the Cliff*. New York: St. Martin's Griffin, 1989.

Menashe, Louis. "Filming Sokurov's Russian Ark: An Interview with Tilman Büttner". *Cineaste* 28, Issue 3 (2003): 21–23.

Merleau-Ponty, Maurice. *The Phenomenology of Perception*. London: Routledge, 1962.

Morrey, Douglas and Dauncey, Hugh. "Quiet Contradictions of Celebrity Zinedine Zidane, Image, Sound, Silence and Fury". *International Journal of Cultural Studies* 11, Issue 3 (2008): 301–318.

Moussavi, Farshid and Zaera-Polo, Alejandro. "Code Remix 2000". In: C. Jencks (ed), *Theories and Manifestos of Contemporary Architecture*. London: Academy Press, 2006, p. 338.

Perez, Gilberto. "Film in Review". *The Yale Review* 89, Issue 1 (2001): 185–193.

Ray, William. *Literary Meaning: From Phenomenology to Deconstruction*. Oxford: Basil Blackwell Ltd., 1984, p. 147.

Schlegel, Hans-Joachim. "Russian Ark: Review". *Film-Dienst* 56, Issue 9 (2003): 22–23.

Spuybroek, Lars. "Machining Architecture". In: C. Jencks (ed), *Theories and Manifestos of Contemporary Architecture*. London: Academy Press, 2006, p. 351.

Tafuri, Manfredo. "Peter Eisenman: Meditations of Icarus". In: Peter Eisenman (ed), *House of Cards*. Oxford: Oxford University Press, 1987, p. 29.

Tschumi, Bernard. *Architecture and Disjunction*. Cambridge, MA: MIT Press, 1996, p. 121.

———. *Architecture and Disjunction*. Cambridge, MA: MIT Press, 1997, p. 122.

———. *Event-Cities*. Cambridge, MA: MIT Press, 1994, p. 11.

———. *Manhattan Transcripts*. London: Academy Editions, 1981.

———. "Parc de la Villette, Paris". In: A. Papadakis, C. Cooke and A. Benjamin (eds), *Deconstruction Omnibus*. New York: Rizzoli, 1989, pp. 175–181.

Whiteman, John. *Investigations in Architecture: Eisenman Studios at GSD: 1983–85*. Cambridge, MA: Harvard University Press, 1986, p. 7.

Wigley, Mark. *The Architecture of Deconstruction: Derrida's Haunt*. Cambridge, MA : MIT Press, 1995.

Wood, Ashley. "Encounters at the Interface: Distributed Attention and Digital Embodiments". *Quarterly Review of Film and Video* 25, Issue 3 (2008): 219–229.

Zaera-Polo, Alejandro (ed). *El Croquis: Peter Eisenman*. Madrid: El Croquis Editorial, 1997.

Zellener, Peter. *Hybrid Space: New Forms in Digital Architecture*. New York: Thames and Hudson, 1999.

Zuckoff, Mitchell. *Robert Altman: The Oral Biography*. London: Vintage, 2010.

9 Intersecting frames
Film + architecture

Scott McQuire

The end of the twentieth century saw a sudden flurry of concern around the themes of film and architecture, or cinema and the city.[1] It is difficult to ignore the coincidence of this surge of attention with the profound transformation that both film and architecture were experiencing, as each practice found itself being increasingly redefined by digital technologies. The much-celebrated centenary of cinema in 1995 also marked a growing uncertainty about what cinema would become in the looming post-celluloid era. Similarly, architecture found itself at a crossroads, as computer-assisted design systems, first mooted in the 1960s, were turbo-boosted by new visualization capabilities. As digitizing photographic and video images became commonplace, the static forms of elevation and plan ceded authority to animated 3D 'fly-throughs', leading Eleftheriades to argue that 'the world of architecture will merge imperceptibly with the world of cinema.'[2] But what would such a merging entail?

In trying to sketch an answer to this question without appealing to a stable essence of either field, this chapter juxtaposes three different moments over the course of the twentieth century in which the relation between film and architecture became the subject of critical attention. Divorced from a narrow definition tied to a specific technical apparatus, it is proposed that 'film' is better recognized as naming the historic threshold in which time and movement become integral to the visual image. Defined in terms of the orchestration of mobile and dynamic fields of vision, 'film' remains not only useful but essential to understanding the new conditions of transparency and opacity, of seeing and being seen, that continue to influence contemporary architecture. From this perspective, film and architecture are both implicated in the production of a new sense of social space born at the junction of increasingly 'open' architectural structures and mobile fields of vision. This is the unstable terrain that today grounds architecture in the digital urban milieu.

Film as urban dynamite

In 1924, Hungarian-born artist László Moholy-Nagy completed the script-collage for a proposed film *Dynamic of a Metropolis*.[3] Never shot, the non-fiction scenario sets out a complex interplay of abstract and documentary elements which not only demonstrates Moholy-Nagy's prodigious capacity to work across varied media but also showcases the extent to which he already saw the camera as the primary tool for the articulation of a new concept of space-time. 'Space creation' (*Raumgestaltung*) through the controlled use of light and movement remained the credo of Moholy-Nagy's work over the next decades, not only in his paintings, lithographs and woodcuts but also in his pioneering work in the era's 'new media' – photomontages, photograms and films – as well as new forms such as the *Lictrequisit* or Light-Space Modulator (1930) that pointed towards kinetic art.

Moholy-Nagy's interest in rendering the *dynamic* qualities of the modern metropolis in film were widely shared in a decade that gave birth to the 'city symphony' film genre. If the most famous products of this cycle were undoubtedly Walter Ruttmann's *Berlin, Symphony of a Great City* (Germany 1927) and Dziga Vertov's *Man with the Movie Camera* (USSR, 1929), its influence stretched much further.[4] In fact, from the very beginning of cinema, it was evident that the *liveliness* of urban phenomena – the restless movement of crowds and vehicles, the shimmering appearance of the modern street under electric lights and reflected in glass structures, the complexity of social experience lived amidst heterogeneous strangers, mass commodities and new media technologies all auspiced by the paradigmatic impact of machine production – were a major fascination for filmmakers. If film offered artists a new medium, the modern city presented a palette of sounds and visions never previously experienced. Novelist Virginia Woolf advanced what was a widely shared view of the early *avant-garde*: that film could become a new art uniquely appropriate to modern urban experience if only it stayed clear of the nostalgic lure of 'literary' scenarios.

> The most fantastic contrasts could be flashed before us with a speed which the writer can only toil after in vain [. . .] We get intimations only in the chaos of the streets, perhaps when some momentary assembly of colour, sound, movement, suggests that here is a scene waiting a new art to be transfixed.[5]

While numerous writers and filmmakers shared this perception of cinema's distinctive relation to the modern city, the most developed theorization of this role emerged in the writings of Walter Benjamin. Beginning in the 1920s, Benjamin advanced a novel conception of film as a form of urban 'dynamite' capable of unlocking the experiential 'prison-world' of the industrial city. The beginnings of this thesis were first proposed in his 1927 defence of Eisenstein's *Battleship Potemkin*, written shortly after Benjamin had visited the Soviet Union.

> To put it in a nutshell, film is the prism in which the spaces of the immediate environment – the spaces in which people live, pursue their avocations, and enjoy their leisure – are laid open before their eyes in a comprehensible, meaningful and passionate way. In themselves these offices, furnished rooms, salons, big-city streets, stations, and factories are ugly, incomprehensible, and hopelessly sad. Or rather, they were and seemed to be, until the advent of film. The cinema then exploded this entire prison-world with its dynamite of fractions of a second, so that now we can take the extended journeys of adventure between their widely scattered ruins. The vicinity of a house, a room, can include dozens of the most unexpected stations, and the most astonishing station names. It is not so much the constant stream of images as the sudden change of place that overcomes a milieu which has resisted every other attempt to unlock its secret, and succeeds in extracting from a petty-bourgeois dwelling the same beauty we admire in an Alfa Romeo. And so far, so good.[6]

Benjamin's argument precociously joins several lines of thought, linking his analysis of the impact of camera technologies on perception to his reading of the impact of 'big-city life' on the human sensorium. When his 'film dynamite' metaphor is reprised a decade later in his famous 'Artwork' essay, it has been ambitiously extended to include the tantalizing concept of the 'optical unconscious' derived from his reading of Freud's *Beyond the Pleasure Principle* (published in 1920).

> Our bars and city streets, our offices and furnished rooms, our railroad stations and factories seemed to close relentlessly around us. Then came film and exploded this prison-world with the dynamite of the split-second, so that we now can set off calmly on journeys of adventure among its far-flung debris. With the close-up, space expands; with slow motion, movement is extended. And just as enlargement not merely clarifies what we see indistinctly 'in any case', but brings to light entirely new structures of matter, slow motion not only reveals familiar aspects of movements, but discloses quite unknown aspects within them [. . .] Clearly, it is another nature which speaks to the camera as compared to the eye. 'Other' above all in the sense that a space informed by human consciousness gives way to a space formed by the unconscious. [. . .] It is through the camera that we first discover the optical unconscious, just as we discover the instinctual unconscious through psychoanalysis.'[7]

Benjamin's argument here has several parts to it. The first, strongly influenced by Baudelaire and Simmel, is his assertion that the quintessential experience of modern city life is a form of shock. Complementing Simmel's thesis concerning the 'blasé attitude' adopted by overburdened city dwellers in order to cope with routine overstimulation,[8] Benjamin used his reading of Freud to argue that consciousness routinely functions as a 'protective shield'. Habituation to shock enables consciousness to 'screen' external stimulation more efficiently, negating shock by locating the stimulus in a unilinear temporal chain. For Benjamin, this made the notionally private life of individual experience a highly political issue. Rather than entering the *durée* of 'long experience', in which individual memory could potentially be articulated with a collective past (and thereby a shared future), he argued that habituation to urban shock had reduced modern social life to isolated and incommunicable individual experiences.

> The greater the shock factor in particular impressions, the more vigilant consciousness has to be in screening stimuli; the more efficiently it does so, the less these impressions enter long experience (*Erfahrung*), and the more they correspond to the concept of isolated experience (*Erlebnis*). Perhaps the special achievement of shock defence is the way it assigns an incident a precise point in time in consciousness, at the cost of the integrity of the incident's contents. This would be a peak achievement of the intellect; it would turn the incident into isolated experience.[9]

Film's 'other space' presents a potential antidote to this isolation. Like architecture, popular reception of film was characterized by what Benjamin termed 'distracted' perception.[10] But film is the 'true training ground' for this mode of perception.[11] Instead of valorizing the kind of focused attention routinely presumed as desirable for viewing art with a capital 'A', the radical value of film lay in its capacity to engender a mode of reception capable of eluding the habitual filters of consciousness.

This understanding situates the historical importance that Benjamin attributed to film as a means of responding to the distinctive historical challenges of 'big-city' life. Slipping past the protective shield of consciousness, Benjamin argued that film was capable of detonating the powerful 'memory traces' that pertained to the sort of experiences which had never properly entered consciousness. It is from this perspective that the full weight of the concept of the 'optical unconscious' can be appreciated: the distracted perception characteristic of film is the key to unlocking the latent historical energy of the modern city, and thereby re-enlivening the utopian aspirations that were embedded in many of its forms and structures.[12] With its capacity to address a collective audience, cinema offered the toolkit that could transform *avant-garde* aesthetics into a genuine political force. In Benjamin's pithy terms, 'The extremely backward attitude to a Picasso painting changes into a highly progressive reaction to a Chaplin film.'[13] Schooled by the 'dynamite of the 1/10th of a second', urban dwellers could perceive their surroundings and their social lives anew and remake them according to their needs.[14]

Of course, the problem for this new politics was how to convert the enervation of distracted perception into conscious awareness, a 'profane illumination' that might be collectively acted on. As Adorno made clear in his critique of Benjamin and Kracauer, he thought the dynamite of montage could only get you so far.[15] If judicious use of such a tool demanded a new type of revolutionary artist-architect, history suggests the fuse is still to be properly lit.

The house as sight machine

In 1954, Alfred Hitchcock's film *Rear Window* famously dramatized the psycho-sexual tensions generated by the convergence of new visual technologies with modern architecture. Hitchcock's film explores the novel capacity for high-powered telephoto lenses to strip the veil of privacy from the glass-windowed houses that had become typical of bourgeois life. *Rear Window* is transgressive, but in a deliberate, stylized way that suggests its transgression is fast becoming normalized. From the film's beginning, it is made clear that the injured photojournalist, Jeff (played by James Stewart and purportedly based on famed photographer and Magnum photo-agency founder Robert Capa), is indulging in 'improper' activities. His nurse, Stella, comments, 'Oh dear, we've become a race of Peeping Toms'. However, by the end of the film, she has become an avid watcher herself. The same is true of Jeff's girlfriend, Lisa (played by Grace Kelly), who is initially disturbed by Jeff's voyeurism, but is soon drawn into watching his neighbor, the suspected murderer, Therbold. More significantly, Hitchcock's film deliberately implicates the film viewer in this uncertain scene, highlighting the extent to which the growing 'openness' of modern architecture and the increasing 'reach' of modern media had converged to produce a new social milieu of urban visibility.[16]

Advocacy of glass architecture was emblematic of modern architectural style and attitude. However, there is a vast difference between the lingering expressionism of a Bruno Taut (member of the Crystal Chain group formed immediately after the First World War) and the enthusiastic adoption of 'window-walls' by arch-rationalist Le Corbusier. History suggests Le Corbusier's aesthetic won the day, but perhaps not with the effects he anticipated. His argument in favour of the horizontal window, first proposed for his Geneva Villa in 1923, is couched entirely in terms of transparency rather than the more diffuse, refracted and colourful light that those such as Taut and Paul Scheerbart (1972) had envisioned. Because horizontal windows provide more light than vertical ones (an argument Le Corbusier 'proves' with reference to data on film exposure), they are better able to bring 'the immensity of the outer world into the room'.[17] This novel form of transport provided by transparency, which is predicated on the availability of modern building materials, ushers in a new definition of the house. By 1930, Le Corbusier was proclaiming, 'With reinforced concrete you get rid of the walls completely. [. . .] [I]f I want to, I can have windows on the entire surface of façade'.[18] Elsewhere he adds: 'From this emerges the true definition of the house: stages of floors [. . .] all around them walls of light'.[19]

As glass takes over the walls of the house, the camera becomes a key point of reference for the new spatial dynamic that is being enacted. In Corbusier's words,

> The house is a box raised above ground, perforated all around, without interruption, by a long horizontal window. [. . .] It is in its right place in the rural landscape of Poissy. But in Biarritz, it would be magnificent. [. . .].[20]

If, as Le Corbusier argues, his new window system is akin to a camera aperture designed to regulate the entry of light, the house (as Colomina concludes) has become 'a system for taking pictures'.[21] The window-wall makes the house a sight machine that can be pointed anywhere, 'a camera pointed at nature', converting landscape into image.[22]

However, as Hitchcock's film suggests, the directionality of the camera lens is always variable. The drama of *Rear Window* turns entirely on the reversible nature of the social visibility that glass architecture enacts. At the very moment that conclusive evidence of a murder having taken place is revealed to both the film's characters and its spectators via a close-up shot putatively seen through photojournalist Jeff's camera lens, the murderer, Therbold, turns to look directly towards the window of Jeff's apartment. This inspires a panicked reaction in Jeff and Lisa ('Turn out the light, he's seen us'). This reaction functions simultaneously on narrative and meta-narrative levels: while it forms part of the fictional story, signaling the moment in which the watchers are seen and thus brought into danger, it also performs a commentary on the way that film watching has helped to normalize voyeurism in an increasingly media-driven society. As a visual apparatus, film created a distinctive social situation that authorized viewers to stare at others with an intensity and persistence that would otherwise be considered improper (something that was particularly radical in the early twentieth century, especially for female viewers). *Rear Window* dramatizes the way that the conjunction of glass architecture and photomedia technologies effectively transformed the private sphere into what Habermas termed 'floodlit privacy'.[23]

Is the modern house a machine for taking pictures, or the studio in which they are taken? As much as modern architecture restructured the house into a light-filled space from which its sovereign inhabitants might survey the world outside, transparency also functioned to render this notionally private space increasingly open to the gaze of others. William Whyte's (1969, 324) influential description of the new living spaces in the post-war suburbs of the United States pinpointed what still remains a staple theme of film and television production in the present: 'The picture in the picture window. . . is what is going on *inside* – or what is going on inside other people's picture windows'. By the 1990s, the spectacle of family life in the home as a source of popular melodrama had escalated into fantasies of 'real-time' omniscience, and the fictional conceit of *The Truman Show* (1998) – in which an entire town become a studio set – found itself outstripped by the global popularity of 'reality television' formats such as *Big Brother*. Here the explosion engineered by the meshing of film and architecture produces what Paul Virilio calls a condition of perpetual *overexposure*.[24]

City as cinematic event

In 1982, Godfrey Reggio's remarkable first feature *Koyaanisqatsi* was released in cinemas, largely thanks to backing from Francis Ford Coppola. *Koyaanisqatsi* reprised the city-symphony genre, particularly as practiced by Vertov, combining a fundamentally musical structure developed in collaboration with composer Phillip Glass with an ambitious non-verbal conception of modern society that spans the underlying processes as well as the sites of urban life. However, where Vertov's *Man with the Movie Camera* presented an optimistic picture of urban-industrial life, Reggio depicts the city-machine in a far more ominous light. Inspired by Ellul's analysis of technological society, *Koyaanisqatsi* shows a city in which human lives are being reduced to mere cogs in a fundamentally non-human system. Reggio acknowledges his focus on systems and structures was a deliberate stylistic choice:

> I try to eradicate all the foreground of traditional film and make the background
> . . . or what's called 'second unit', the foreground . . . I was trying to look at
> buildings, masses of people, transportation, industrialization, as *entities* in and of
> themselves, having an autonomous nature.[25]

Perhaps the most striking outcome of this approach is the remarkable section of the film sub-titled 'The Grid'. Director of photography Ron Fricke exploits the capacity of time-lapse cinematography to depict urban life on the move. The result is a memorable, extended sequence that fulfills Benjamin's observations about camera technologies providing access to the 'optical unconscious'. Urban life flies by at high speed, but this rapid-seeing – perhaps paradoxically – enables new sets of rhythms to be apprehended: the frenzied, Brownian motion of pedestrians is revealed as staccato bursts of activity metred by the guillotine of traffic signals; the slow pulse snaking up a line of cars on a freeway forms a standing wave in response to a momentary hold-up that occurred far away in space and time; the night city under conditions of acceleration dissolves structure and mass into the uncanny beauty of streaks of pure light. Reggio's film confirms Benjamin's hypothesis of film as a form of 'urban dynamite'. It reveals previously unperceived patterns of city life and helps to crystallizes a different understanding of architecture in which the primary concern is neither the formal appearances or spatial envelope of static structures but the processes and flows that occur across space-time.

At least since Futurism, modern architecture has evinced an interest in the dynamic and ephemeral aspects of urban life, a tendency that undoubtedly strengthened its perceived affinity with cinema. Arguably, the most significant statement of the importance of flow emerged in the work of Sigfried Giedion, who was strongly influenced by his close friend Moholy-Nagy's concept of 'space creation'. In his influential book *Space, Time and Architecture*, which became a staple of architectural education for decades, Giedion contrasts the aesthetics of previous eras with the 'space-time' of the modern city.[26] One prominent sign for Giedion was the blurring of interior and exterior space enabled by glass architecture. In his optimistic reading of Walter Gropius' landmark Bauhaus building erected in Dessau in 1925, Giedion argues that Gropius' startling use of glass curtain walls presaged 'an epochal move away from Renaissance spatiality'.[27] This technological threshold, predicated on the application of new construction materials including steel girders and reinforced concrete, carried perspectival ambiguity into new sites and situations. Glass structures lessened the dominion of a stable, centred perspective in favour of a more dynamic and relational point of view. While Giedion's account draws on examples from modern art (such as Picasso) and architecture, it also – and perhaps more tellingly for this argument – draws on the impact of modern transport infrastructure. Seizing the new 'parkway' outside New York as an exemplar, Giedion argued,

> As with many of the creations born out of the spirit of this age, the meaning and beauty of the parkway cannot be grasped from a single point of observation, as was possible from a window of the château at Versailles. It can be revealed only by movement, by going along in a steady flow, as the rules of traffic prescribe. The space-time feeling of our period can seldom be felt so keenly as when driving.[28]

As Paul Virilio has noted, driving and cinema share a common perceptual frame.[29] Giedion's observation about the challenge of representing modern design indicates a similar convergence between moving vehicle and cinema's mobile vision: "Still photography does not capture them clearly. One would have to accompany the eye as it moves: only film can make the new architecture intelligible".[30]

Such a standpoint situates a paradigmatic change in thinking about architecture. Filmmaker and theorist Hollis Frampton has argued, 'Painting "assumes" architecture: walls, floors, ceilings.[31] The illusionist painting itself may be seen as a window or doorway'. By contrast, cinema's dynamic perception – 'perception in the form of shocks' as Benjamin put it – 'assumes' not the stable site of a solid building but the variable vector described by a moving vehicle. If Frampton's evocation of 'architecture' remains classical, Virilio's postulation of the unity of travel and tracking shots concentrates the essential ambiguity conditioning both modern perception and modern architecture: the endless fluctuation of borders and contexts (inside/outside, public/private, local/global), and the displacement of the human body as authoritative centre of meaning and reference.

The attempt to render architecture more 'dynamic' has assumed numerous forms in modernity. The architectural promenades made famous by Le Corbusier in projects such as *Villa Savoye* – which Beatriz Colomina aptly describes as 'cinematographic' in its choreography of inhabitant movement – gave way to new schemes for producing entire 'cities on the move'. While those such as Constant and Yona Friedman advocated a radical form of urban nomadism based on designing mutable structures that could be reconfigured by inhabitants according to their needs, others such as Archigram and Cedric Price looked to the new potential for computer control to enable new levels of personalization and customization in architecture. A common point of reference for many of these utopian schemes was their appeal to cinema, less as an existing institutional form of mass entertainment than in its potential for generating large-scale spatial ambiences capable of rapid switches and differentiated intensities.

Beginning in the 1970s, Bernard Tschumi adopts cinema as the key reference point for asserting a new relation between architecture and event. In his 1983 essay 'Space and Event', Tschumi argues that where architectural photography tends to reduce architecture to 'a passive "object" of contemplation instead of the *place* that confronts spaces and actions', and contends that any new attitude to architecture needed to question its mode of 'representation'.[32] In some respects, Tschumi repeats Giedion's problematic, while replacing its overweening formalism with a sharper political critique, calling for 'cinematic devices' to replace 'conventional description'.[33] However, his description of the *Manhattan Transcripts* as embodying a broadly filmic conception of architecture is based less on the disruption of the dominant visual regime than the re-insertion of *time* into architecture:

> The temporality of the *Transcripts* inevitably suggests the analogy of film. In both, spaces are not only composed but also developed from shot to shot so that the final meaning of each shot depends on its context.[34]

Tschumi's recourse to the metaphor of film supports the overarching argument that remains on the web archive of the *Manhattan Transcripts*: 'Architecture is not simply about space and form, but also about event, action, and what happens in space'.[35] While this broad understanding has clearly gained ground in the present, the status of the filmic analogy has become more uncertain. Is 'film' a tool to be wielded by a creative practitioner in the service of their distinctive vision? Or is it indicative of a new, more fluid set of social relations of space and time? Where Le Corbusier and Giedion imagine the architect controlling perspective and sequence with the plastic facility of a film director, following what might be called a traditional *authorial* perspective, it is notable that Benjamin compares film and architecture largely in terms of reception. Tschumi's problematic remains striking inasmuch as it is an uneasy mixture of both: grounded in production (the architect as author using filmic tools or notations) its logic points towards reception in which architecture becomes 'event'.

Architecture + film after cinema

When the relation between film and architecture resurfaced as a problematic in the context of the surging development of digital imaging in the 1990s, it was notable that it was not Benjamin's problematic of distracted collective reception which came to the fore, but the problematic of the architect as author – or filmmaker. Once digital imaging, including CAD systems, became more sophisticated as computers gained in storage capacity and processing speed, a number of displacements occurred. Architects were increasingly able to build proto-filmic 'virtual' environments using control of perspective, sequence and duration to construct time-based visual presentations in the mode of film directors. (This was the context of Michael Eleftheriades speculation about a pending convergence between film and architecture with which I began this chapter.) The dominance of this orientation created a paradox. Architects seemed fascinated by the computer as a toolbox for generating sophisticated digital imagery, but remained largely uninterested in its potential for supporting distributed communications. This stance completely reversed the earlier, rather precocious, ambitions articulated by those including Archigram and Nicholas Negroponte in relation to architecture and computing.[36]

At the same time that architecture sought to appropriate a certain ideal of cinematic representation, the authority of such a model was diminishing elsewhere. While moves towards forms of audience-directed cinema in the 1990s largely proved to be dead ends, participatory practices gained ground in other areas, notably in contemporary art as well as new areas such as online gaming. In place of the traditional focus on discrete and finished objects or images came an emphasis on what Umberto Eco aptly called 'open works', describing forms of creative practice that deliberately left room for the audience or for chance.[37]

One of the challenges facing modern architects – and arguably one that underpinned the attractions of film as a model – is the problem of spatial order, particularly the ordering of reception. How do you ensure that the public understands – and utilizes – a building as you intended? In the past, architects might have been able to rely on a shared and relatively stable symbolic order that replicated existing and long-standing social functions and hierarchies. However, as Lefebvre has argued, that shared sense of space and social order was 'shattered' in modernity.[38] In place of a common set of reference points sits the absence of a unified and unifying narrative. This is simultaneously a mark of new freedom – the possibility of reinventing the city and its social life that tantalizes so many utopic architectural programs – and coercion: the fact that we have no choice but to participate in this vast and uncertain social experiment.

The response of many modern architects to the new conditions was often a similar mixture in which broad aspirations for 'freedom' were interspersed with coercive elements and programs. Here we might recall Mies' demand that tenants in his Lake Shore Drive apartments in Chicago only be permitted to install neutral grey curtains lest they mar the external appearances of the building. While such a demand is, in some respects, minor, it is emblematic of the extent to which many architects sought to use design to prescribe the ways in which people could inhabit a building or precinct. It would be simplistic to argue that these attempts either succeeded or failed. If inhabitants often adapted their surroundings to remake the everyday environment and frustrate architectural 'direction', they did not do so in circumstances of their own choosing and usually lacked control over key features.

Despite the appeal of film to architects as a means of ordering the sequence of reception, most filmmakers have experienced similar limits in their capacity to orchestrate meaning and reception. Reggio designed *Koyaanisqatsi* in order to critique contemporary urban-industrial culture as exemplifying 'life out of balance'. However, the techniques he used, particularly the time-lapse photography of the cityscape, has subsequently become one of the most common tropes for depicting the contemporary city for all kinds of purposes and have been widely used in news promos, contemporary dramas and advertising. While arguing that he wanted to demonstrate 'the beauty *of* the beast', Reggio also acknowledges that working in a visual medium like film brings an inevitable loss of 'authorial' control compared to writing:

> What you give up is the specificity of one thought, one idea, unmistakably getting your point across, which people can agree or disagree with. But what you do get is the richness of an experience that can stay in the conscious and unconscious mind and can be continually revisited.[39]

The digital milieu brings a further change to consider in developing a new conception of both film and architecture. Historically, while a filmmaker could capture and display unexpected phenomena, each shot, and its place in the film's final order, had to be pre-selected. While the self-reflexive conceit of a film such as *Man with the Movie Camera* is that the audience 'sees' the footage shot, edited and finally viewed, the film's order had to be fixed in advance. However, the advent of digital databases means this is no longer the only model. Perry Bard's reworking of Vertov's film demonstrates the potential for contemporary film to be reconstituted as an 'open work' in which different manifestations or variations do not exhaust the conception.[40] Today we need to ask can digital infrastructure play a similar role in rendering the social life of built structures similarly mutable and reconfigurable according to user-input and inhabitant desire?

This question situates the persistence and pertinence of the film metaphor for architecture. Even as 'film' relinquishes its dependence on celluloid and sprockets in favour of digital files and LED illumination, time-based, mobile fields of vision remain central in defining contemporary urban experience. If the multiplicity of perspectives, the fusion or confusion of the boundary between interior and exterior enacted by the glass window-wall and the profusion of ephemeral structures in the present all speak to a *cinematic* disturbance at the heart of modern architecture, this encompasses the need to rethink the concept of space to incorporate practices and processes – all those temporal phenomena that define the city in terms of what Tschumi called event-space.

However, this cannot be a simple recapitulation of earlier programs. Benjamin's conception of film as urban dynamite, like Reggio's *Koyaanisqatsi*, point to what Marshall McLuhan – drawing on *gestalt* psychology as much as cybernetics – long ago placed at the centre of his own analysis: *pattern recognition*.[41] If the modern city is an event in which the authority of centred perception gives way to perception in motion, how do we gain access to it and develop an understanding of its dynamics? The kind of cinematic response that *Koyaanisqatsi* offered to the complexity of the modern city is fast being overtaken by new forms of data visualization irrigated by fast-flowing streams of sensor-based information subject to high-speed algorithmic analysis. Is the 'smart city' a new version of the authorial or directorial control that film once seemed to offer to architecture? Where, in these data-rich scenarios with their emphasis on control over the future city, is the space for the eruptive and interruptive nature of the 'event' naming the as-yet unassimilated and unassimilable happening – which might also be a name for the action and activity of the city's inhabitants, the unthought or after-thought in so many heroic architectural scenarios?

Notes

1 'Film and architecture' approaches tended to emerge from architectural circles. See Albrecht (1986), Toy (1994), Neumann (1996), Fear (2000), and Lamster (2000). For 'cinema and the city', which tended to emerge from film and cultural studies, see Clarke (1997), Penz and Thomas (1997), Shiel and Fitzmaurice (2001), and Barber (2002).

2 Michael Eleftheriades, 'Architecture or Cinema: Digital 3D Design and the World of Multimedia,' in *Cinema and Architecture: Méliès, Mallet-Stevens, Multimedia*, ed. François Penz and Maureen Thomas (London: BFI, 1997), 143.

3 The text later was translated into German and published in 1925 as an appendix to *Painting Photography Film*, the eighth volume in the Bauhaus book series that Moholy-Nagy directed.

4 I began with Moholy-Nagy's script because he started writing it in 1921, the same year Paul Strand and Charles Sheeler made their Whitmanesque short film *Manhatta* in New York. Other films in the 'city symphony' cycle in the 1920s include René Clair's *Paris qui dort* (1923) and *Entr'acte* (1924), Alberto Cavalcanti's *Rien que les heures* (France, 1926), Joris Iven's *De Brug* (Holland, 1928) and *Regan* (Holland, 1929) and Jean Vigo's *A propos de Nice* (France, 1930). The genre continues to reappear periodically: for example Arne Sucksdorff's *Människor i stad* (Symphony of a city, USA/Sweden, 1948), Francis Thompson's *N.Y.,*

N.Y (USA, 1958), Godfrey Reggio's *Koyaanisqatsi* (1982) and Hubertus Siegert's *Berlin Babylon* (Germany 1996–2001).

5 Virginia Woolf, 'The Cinema,' in *Collected Essays* vol. 2 (London: Hogarth Press, 1966), 272.

6 Walter Benjamin, *Selected Writings, Vol. 2, 1927–1934*, ed. M. Jennings, H. Eiland and G. Smith, trans. R. Livingstone and others (Cambridge, MA: Belknap Press, 1999), 17.

7 Walter Benjamin, *Selected Writings, Vol. 3, 1935–1938*, ed. H. Eiland and M. Jennings, trans. E. Jephcott, H. Eiland and others (Cambridge, MA: Belknap Press, 2002), 117.

8 Georg Simmel, *Simmel on Culture*, ed. D. Frisby and M. Featherstone (London: Sage, 1997), 178.

9 Walter Benjamin, *Selected Writings, Vol. 4, 1938–1940*, ed. H. Eiland and M. Jennings, trans. E. Jephcott and others (Cambridge, MA: Belknap Press, 2003), 319.

10 Here he follows Siegfried Kracauer's 1926 essay 'Cult of Distraction'.

11 Benjamin, *Selected Writings, Vol. 4*, 269.

12 Here Benjamin was strongly influenced by surrealism. In his essay on surrealism, Benjamin argued that André Breton was 'the first to perceive the revolutionary energies that appear in the 'outmoded' – in the first iron constructions, the first factory buildings, the earliest photo, objects that have begun to be extinct [. . .] No one before these visionaries and augurs perceived how destitution not only social but architectonic, the poverty of interiors, enslaved and enslaving objects – can suddenly be transformed into revolutionary nihilism. [. . .] They bring the immense forces of 'atmosphere' concealed in things to the point of explosion' (Benjamin, *Selected Writings, Vol. 2*, 210).

13 Benjamin, *Selected Writings, Vol. 3*, 116.

14 Benjamin, *Selected Writings, Vol. 4*, 264.

15 Theodor W. Adorno, *Aesthetic Theory* (London: Routledge & K. Paul, 1984), 223.

16 Jacques Tati's *Playtime* (1967) replays this scenario, but this time as farce. J.G. Ballard's story *The 60 minute Zoom* (1976) also explores the complex emotional terrain of the close-up, but here the voyeur watching his wife's infidelity becomes the murderer.

17 Le Corbusier cited in Bruno Reichlin, 'The Pros and Cons of the Horizontal Window,' *Daidalos* 13 (1984): 72.

18 Le Corbusier, *Precisions*, trans. ES Aujame (Cambridge, MA: MIT Press, 1991), 38–40.

19 Le Corbusier cited in Beatriz Colomina, *Publicity and Privacy: Modern Architecture as Mass Media* (Cambridge, MA: MIT Press, 1994), 7.

20 Le Corbusier, *Precisions*, 136, 139.

21 Colomina, *Publicity and Privacy*, 312.

22 Ibid.

23 Jürgen Habermas, *Structural Transformation of the Public Sphere: An Inquiry into a Category of Bourgeois Society*, trans. T Burger with the assistance of F. Lawrence (Cambridge, MA: MIT Press, 1989), 158.

24 Paul Virilio, *The Information Bomb*, trans. C. Turner (London: Verso, 2000), 60–1.

25 Godfrey Reggio cited in Scott MacDonald, *A Critical Cinema: Interviews with Independent Filmmakers* (vol. 2) (Berkley: University of California Press, 1992), 390.

26 Siegfried Giedion, *Space, Time and Architecture: The Growth of a New Tradition* (Cambridge, MA: Harvard University Press, 1967).

27 Siegfried Giedion cited in Detlef Mertins, 'Anything but Literal: Sigfried Giedion and the Reception of Cubism in Germany,' in *Architecture and Cubism*, ed. E. Blau and N. Troy (Montréal: Canadian Centre for Architecture, 1997), 234.

28 Giedion, *Space, Time and Architecture*, 826, 831.

29 Paul Virilio, *War and Cinema: The Logistics of Perception*, trans. P. Camiller (London: Verso, 1989).

30 Siegfried Giedion cited in Andres Janser, '"Only Film Can Make the New Architecture Intelligible!" Hans Richter's Die Neue Wohnung and the Early Documentary Film on Modern Architecture,' in *Cinema and Architecture: Melies, Mallet-Stevens, Multimedia*, ed. François Penz and Maureen Thomas (London: BFI, 1997), 34.

31 Hollis Frampton, *Circles of Confusion* (New York: Visual Studies Workshop Press, 1983), 189.

32 Bernard Tschumi, 'Spaces and Events,' in *Questions of Space: Lectures on Architecture* (London: AA Publications, 1990), 89.

33 Tschumi, 'Spaces and Events,' 90, 95.
34 Bernard Tschumi, 'Index of Architecture: Themes from The Manhattan Transcripts,' in *Questions of Space: Lectures on Architecture* (London: AA Publications, 1990), 107.
35 http://www.tschumi.com/projects/18/ [Accessed 16 November 2015].
36 Scott McQuire, *The Media City: Media, Architecture and Urban Space* (London: Sage, 2008).
37 Umberto Eco, *The Open Work*, trans. A. Cancogni (Cambridge, MA: Harvard University Press, 1989).
38 Henri Lefebvre, *The Production of Space*, trans. D. Nicholson-Smith (Oxford: Blackwell, 1991), 25.
39 Reggio in MacDonald, *A Critical Cinema*, 389.
40 Bard's *Man with a Movie Camera: Global Remake* project, begun in 2005, enables people to contribute image sequences 'interpreting' Vertov's original film. It uses specially developed software to archive, sequence and stream the different contributions as a 'participatory film'. See http://dziga.perrybard.net/
41 Recall the famous statement in *Understanding Media* (1974, 8): 'The "message" of any medium or technology is the change of scale or pace or pattern that it introduces into human affairs', while in *Counterblast* (1970, 132) he argues, 'Faced with information overload, we have no alternative but pattern-recognition.' Elsewhere McLuhan declared, 'I am a pattern watcher. (Marshall McLuhan and David Carson, *The Book of Probes*. ed. Eric McLuhan and William Kuhns (Berkeley, CA: Gingko Press, 2011), 311.

Bibliography

Adorno, Theodor W. *Aesthetic Theory*. London and Boston: Routledge & K. Paul, 1984.

Albrecht, Donald. *Designing Dreams: Modern Architecture in the Movies*. New York: Harper & Row in collaboration with the Museum of Modern Art, 1986.

Barber, Stephen. *Projected Cities: Cinema and Urban Space*. London: Reaktion, 2002.

Benjamin, Walter. *Selected Writings, Vol. 2, 1927–1934*. Edited by M. Jennings, H. Eiland and G. Smith; translated by R. Livingstone and others. Cambridge, MA: Belknap Press, 1999.

Benjamin, Walter. *Selected Writings, Vol. 3, 1935–1938*. Edited by H. Eiland and M.W. Jennings; translated by E. Jephcott and others. Cambridge, MA: Belknap Press, 2002.

Benjamin, Walter. *Selected Writings, Vol. 4, 1938–1940*. Edited by H. Eiland and M.W. Jennings; translated by E. Jephcott and others. Cambridge, MA: Belknap Press, 2003.

Clarke, David B., ed. *The Cinematic City*. London: Routledge, 1997.

Colomina, Beatriz. *Publicity and Privacy: Modern Architecture as Mass Media*, Cambridge, MA: MIT Press, 1994.

Eco, Umberto. *The Open Work*. Translated by A. Cancogni. Cambridge, MA: Harvard University Press, 1989.

Eleftheriades, Michael. 'Architecture or Cinema: Digital 3D Design and the World of Multimedia.' In *Cinema and Architecture: Méliès, Mallet-Stevens, Multimedia*, edited by François Penz and Maureen Thomas, London: BFI, 1997, pp. 138–143.

Fear, Bob, ed. *Architecture + Film, II*. London: Wiley-Academy, 2000.

Frampton, Hollis. *Circles of Confusion*. New York: Visual Studies Workshop Press, 1983.

Giedion, Sigfried. *Space, Time and Architecture: The Growth of a New Tradition*. Cambridge, MA: Harvard University Press, 1967.

Habermas, Jürgen. *Structural Transformation of the Public Sphere: An Inquiry into a Category of Bourgeois Society*. Translated by T. Burger (with the assistance of F. Lawrence). Cambridge, MA: MIT Press, 1989.

Janser, Andres. '"Only Film Can Make the New Architecture Intelligible!" Hans Richter's *Die Neue Wohnung* and the Early Documentary Film on Modern Architecture.' In *Cinema and Architecture: Melies, Mallet-Stevens, Multimedia*, edited by François Penz and Maureen Thomas. London: BFI, 1997, pp. 34–49.

Lamster, Mark, ed. *Architecture and Film*. New York: Princeton Architectural Press, 2000.

Le Corbusier. *Precisions*. Translated by ES Aujame. Cambridge, MA: MIT Press, 1991 (first published 1930).

Lefebvre, Henri. *The Production of Space*. Translated by D. Nicholson-Smith. Oxford: Blackwell, 1991.

MacDonald, Scott. *A Critical Cinema: Interviews with Independent Filmmakers* (vol. 2). Berkley: University of California Press, 1992.

McLuhan, Marshall. *Counter Blast*. London: Rapp and Whiting, 1970.

McLuhan, Marshall. *Understanding Media: The Extensions of Man*. London: Abacus, 1974.

McLuhan, Marshall and Carson, David. *The Book of Probes*. Edited by Eric McLuhan and William Kuhns. Berekley, CA: Gingko Press, 2011.

McQuire, Scott. *The Media City: Media, Architecture and Urban Space*. London: Sage, 2008.

Mertins, Detlef. 'Anything but Literal: Sigfried Giedion and the Reception of Cubism in Germany.' In *Architecture and Cubism*, edited by E. Blau and N. Troy. Montréal: Canadian Centre for Architecture, 1997, pp. 219–251.

Neumann, Dietrich, ed. *Film Architecture: Set Designs from Metropolis to Blade Runner*. New York: Prestel-Verlag, 1996.

Penz, François and Thomas, Maureen, eds. *Cinema and Architecture: Melies, Mallet-Stevens, Multimedia*. London: BFI, 1997.

Reichlin, Bruno. 'The Pros and Cons of the Horizontal Window,' *Daidalos* 13 (1984): 65–78.

Scheerbart, Paul. *Glass Architecture*. Translated by J. Palmes. New York: Praeger, 1972.

Shiel, Mark and Fitzmaurice, Tony, eds. *Cinema and the City: Film and Urban Societies in a Global Context*. Oxford: Blackwell Publishing, 2001.

Simmel, Georg. *Simmel on Culture*. Edited by D. Frisby and M. Featherstone. London: Sage, 1997.

Toy, Maggie, ed. *Architecture & Film*. London: Academy Editions, 1994.

Tschumi, Bernard. 'Index of Architecture: Themes from the Manhattan Transcripts.' In *Questions of Space: Lectures on Architecture*. London: AA Publications, 1990, pp. 97–110.

Tschumi, Bernard. 'Spaces and Events.' In *Questions of Space: Lectures on Architecture*. London: AA Publications, 1990, pp. 87–96.

Virilio, Paul. *War and Cinema: The Logistics of Perception*. Translated by P Camiller. London: Verso, 1989.

Virilio, Paul. *The Information Bomb*. Translated by C. Turner. London: Verso, 2000.

Woolf, Virginia. 'The Cinema.' In *Collected Essays*. vol 2, London: Hogarth Press, 1966, pp. 268–272.

Part IV
Digital technologies

10 Hyperreality, vision and architecture

David Ross Scheer

Introduction: The shock of the not-so-new

Architecture is undergoing epochal changes as drawing is replaced by digital design technologies. These technologies, chiefly building information modeling and computational design, are rapidly becoming the chief media of architectural design and communication. The salient characteristics of these technologies are (a) their use of a "three-dimensional" digital model as the design interface and (b) their incorporation of design parameters and building data in addition to building geometry. Such models are simulations, not only in the visual sense but also in the larger sense that they replace reality with its operational equivalent.[1] As simulations, such models both demand and produce in their users a new mode of visual perception. This has given rise to a positive feedback loop of advancements in the technology and increased reliance on the simulations it creates. Architecture is reaching a point of no return. Representational modes of seeing and thinking are going extinct and are now found primarily among an aging generation of architects whose activity and influence are on the wane. Architects are increasingly preoccupied with pre-experiencing their designs in realistic renderings and animations. They experiment with immersive environments and other types of virtual reality and eagerly adopt each new development of such technologies to endow their pre-experience with ever-greater degrees of realism. This elision of image and experience has profound consequences for architectural design and ideation.

Hyperreality, operationalism and simulation

As first described by Jean Baudrillard,[2] hyperreality is a condition in which meaning is replaced by *operation*: a thing or idea is defined by what it *does* rather than what it represents. Hyperreality admits only operational understanding and cannot provide any knowledge about means or causes.[3] This sufficiency of operational knowledge emerges from the modern Western *epistemē* in which having knowledge of something means being able to account for its origins in an unbroken chain of cause and effect.[4] Once the processes that give rise to a phenomenon are understood, they can be manipulated to exert control over it. Science (to call this type of knowledge by its name) does not engage questions of meaning, only of efficient cause. The technology that flows from it naturally embodies the same values.[5]

Baudrillard used the term *simulation* to describe the workings of hyperreality. In popular usage, a "simulation" is an experience that "feels real," produced by artificial means, usually (but not necessarily) digital. In the former sense, simulation is a pervasive cultural phenomenon that undermines reality itself. In the latter, it is a limited, circumscribed experience detached from a background that is perceived as reality. Nevertheless, the two meanings are related in their shared operational basis. The user of a simulation (in the popular sense) implicitly agrees to overlook the obvious fact that it is produced by radically different means than was the experience it evokes. In other words, the user accepts the experience as given without reference to the processes that produced it. What matters is the experience itself, not how it was produced. The operational aspect of reality – its experiential effects – is the entire content of simulation in both senses. Anything in reality that we do not observe has no effect on our experience and may be omitted from a simulation.

Representation is the predecessor and complement of simulation. It preserves the reality principle, i.e., that there is a world that exists independently of our experience of it that always exceeds the ideas we construct of it. Representation relies on signs that refer to objects in the real world. All signs leave gaps and ambiguities in their relationship to reality. Our awareness of these inevitable disparities contributes to the creation of meaning by stimulating exploration of parts of our experience that have thus far eluded representation. From this perspective, a simulation is a fraudulent representation in that it claims to completely reproduce a part of the world and is fraudulent because all representations are partial. From the perspective of simulation, representation is eliminated, made over as simulation. In simulation, no external references exist – it is a WYSIWYG (what-you-see-is-what-you-get) world. Simulation is based on an entirely different principle: rather than acknowledging and exploiting the difference between the sign and reality, *it replaces reality with its sign*. Simulation equates immediate experience with the operation of reality.[6,7]

Hyperreality can be understood as applying the cognitive orientation of the user of a limited simulation to experience in general. In hyperreality, all experience is received at face value without reference to how it was produced. Simulation has the potential to contaminate all of reality with its refusal of depth and reference. Simulation is viral: it can convert whatever it touches to simulation, like ice-nine in Kurt Vonnegut's novel *Cat's Cradle*.[8] Once the identity of operation and reality is established, there is no stopping point. There is no putative reality to return to. Thus universalized, simulation doesn't merely deprive signs of their referents, it eradicates their very possibility – the very idea of representation. A culture immersed in hyperreality cannot conceive of an environment that signifies anything because it no longer understands signification.

Hyperreal visuality

The chief characteristic of hyperreal visuality is the pervasive elimination of the difference between image and reality. It is its pervasive nature, rather than a mere tendency to experience images as realistic, that is determinative. This tendency appears to be an aspect of human perception. Fidelity to nature has been a valued attribute of many types of visual representation. When they were the best medium for capturing a person's appearance, portrait paintings were valued for their resemblance to their subjects. The painted panoramas exhibited in eighteenth-century Paris and London were popular because of the experience they gave viewers of being in a different place or time.[9] When photography was first introduced, the public was fascinated by the immediacy of its images.[10] By today's standards, of course, early photographs hardly seem likely to be taken as bits of reality. But the obvious ways in which early photographs do not resemble actual vision did not prevent their reception as "realistic". As these examples show, the impulse to experience images as "real" has little to do with fidelity per se. Viewed objectively, a modern digital photograph is not much more realistic than a daguerreotype. It may be clearer and show color, but it is still a two-dimensional, framed excerpt of a visual scene. Only our seemingly inborn inclination to see images as reflections of reality allows us to endow any of these images with any degree of realism.

The differences between reality and mechanically produced images did not pass unnoticed. Critical practitioners of photography and cinema began to challenge the inclination to overlook these differences and show their images for what they are: representations that, while providing a view of some part of reality, are also expressions of the photographer's intentions.[11] The distinctive artistic potential of these media was thus established. In popular media, by contrast, producers usually do their best to immerse viewers in the media experience, creating simulations whose self-contained worlds do not call for – indeed, could not survive – any direct contact with external reality. As popular media, photography and cinema demand a visuality that overlooks the many ways their images depart from reality. By making such images a central part of everyday life, television crucially contributed to making this mode of visuality universal and to making hyperreality the near-universal mode of experience in Western culture.[12,13]

Whereas photography, cinema and television at times self-critically examine their manipulation of reality (and can therefore serve as means of enhancing our understanding of it), more recent image technologies have *as their explicit aim* the elimination of the distinction between reality and artifice. Technicolor, CinemaScope™, computer-generated images (CGI), RealD3D™ and virtual reality environments such as the New York Times' recently released NYT VR™ platform[14] each represent a step in this process, demolishing representation and enshrining hyperreality as our culture's mode of visuality and experience in general.[15]

Architectural visual media have undergone a parallel evolution, with one important difference: whereas photographs, movies and television shows can be products to be consumed for their own sakes, the visual artifacts of the architectural design process are broadly understood to refer to a potential building.[16] This provides additional incentive for both architects and clients to keep the differences between artifact and reality/building clearly in mind. Until recently, this was reinforced by the clearly representational nature of drawing. It takes a considerable act of imagination to extrapolate the experience of even a good manually rendered perspective drawing to one of a real building. However, digitally produced images (realistic computer-generated static images, animations and virtual-reality environments) now allow architects and their clients to believe they are seeing projects as they will appear at completion. The reliance of other visual media on hyperreality and simulation makes the reception of these architectural images as reality all but inevitable.

How does the operational nature of hyperreality manifest itself in vision specifically? Obviously, hyperreal visuality must contribute to the experience of simulation described earlier. Since vision is the primary sense through which people gather information about their world (in Western culture at least), visuality must play a major role in hyperreality. People can become immersed in a simulated environment that lacks every sensory stimulus but the visual. An operational visuality by definition would consist of only those aspects of visual experience that produce sensible effects present to any viewer with normal vision. Hyperreal visuality is thus a visuality of direct, intersubjective sensory experience – a visuality of *appearance*. Like all phenomena producing an experience in simulation, those giving rise to visual appearance are irrelevant.

As a result, hyperreal visuality minimizes or excludes interpretations that are necessary to attribute meaning – that is, personal and culturally based associations – to visual experience. That vision occurs without such associations and the cultural bias they entail runs counter to post-structuralist and feminist theories of vision which have found favor with many architectural theorists.[17] It is a common theme in these theories that vision is conditioned by cultural (including gender-based) biases creating varied ways of seeing, rendering universal statements about vision impossible. On the other hand, phenomenologists such as Maurice Merleau-Ponty have maintained that we experience vision prior to any interpretation of it.[18] Furthermore, recent research in the psychology and neuroscience of vision lends credence to the notion that there is a stage of vision that is purely a product of the functioning of the human brain and sense organs.[19] The picture of vision that emerges is a two-stage process: first, we experience so-called early vision, which is a product of human physiology and is thus the same for most people regardless of time or place. In the second stage, objects of vision acquire meanings or interpretations based on associations arising from the observer's personal experience, which is influenced to a considerable extent by the culture in which one has been raised. In these terms, hyperreal visuality elevates early vision and suppresses subsequent interpretation.

The consequences of hyperreal visuality for architecture are transformational, both for its design and its reception. Drawing, with its representational nature, has been the foundation of architectural imagination and communication since the Renaissance. The immersion of contemporary architects in hyperreal visuality radically alters their relationship to the visual artifacts used in design and thereby to the object being designed. Hyperreality and simulation also inevitably affect how works of architecture are understood by the public. These topics will be explored in the next section.

Hyperreal visuality: Architectural design

Architectural design was insulated to a considerable degree from hyperreal visuality by the use of drawings as its chief medium of thought and communication. Architects are traditionally adept at moving back and forth between three-dimensional mental images and two-dimensional representations of a building. There is no possibility of conflating image and reality under these conditions.

Although the impulse to mimic reality was present in architectural representation as in other media, the techniques for achieving it through the twentieth century differed little from those of the eighteenth century. The renderings and models twentieth century architects made to give themselves and their clients "experiences" of completed buildings were in fact often inferior to those of two centuries earlier. Crucially, for the architect's position vis-à-vis the client and the project, the acknowledged incompleteness of these representations allowed the design itself to remain distinct from its representations – an idea in the architect's mind. The architect was thus established as the sole author of the design, and this provided the basis of the architect's authority over the project. In all other respects, however, architects and their clients of this period lived in a culture steeped in hyperreality. They were ready for a new generation of tools that finally admitted hyperreal visuality into architectural design in the first decade of the twenty-first century.

The digital models that architects increasingly use in design are instances of simulation in both the popular and pan-cultural senses. In the popular sense, they are simulations that duplicate the performance of many aspects of a building. This can include not only physical properties, such as structural integrity, energy use and thermal comfort, but also construction processes, such as trade coordination, cost control and scheduling. To perform these simulations, a model must contain as much information as possible about a building's construction so that it ultimately becomes a virtual building – ideally, the building's operational equivalent. These models thus become instances of simulation in the pan-cultural sense – a part of hyperreality. It is more accurate to consider the building as an instantiation of the model than to view the model as a prefiguration of the building.[20]

This conclusion may seem rather perverse. Of course anyone can easily distinguish a model displayed on a computer monitor from a building that occupies space.[21] However, at issue here is *design* – specifically, the relationship among architect, model and building. What is decisive in this context is *the attitude of the architect towards the model*. What does the architect see when looking at a model display? At any given moment it will be one of two things: either a flat surface with a (possibly moving) picture on it, or a window into a world that becomes the architect's present reality. In the former case, the architect is thinking representationally and looking at the model in the same way as one would look at a drawing. In the latter, the architect has crossed over into hyperreality. The latter is much more likely, both because as a culture we have become conditioned to hyperreality in other aspects of our lives and because the models are most useful when viewed as simulation. They allow the user to see the project in perspective, move through it, query any element and find out what it is and garner other information. The model is increasingly the place where architects communicate with colleagues, see others' work and show their own. The model becomes the *topos* where the design team works. In Hannah Arendt's terms, the model is the space where the architect appears to her colleagues and they to her.[22]

Given the current state of modeling technology, the operational equivalence of model and building is far from fully realized. This does not alter the fact that such models demand to be taken as simulations (in both the popular and pan-cultural senses) and their users are strongly influenced to experience them as such. Once again, it is the attitude of the designer that matters. A low level of "realism" does not necessarily diminish the experience of simulation. One can see this in computer games such as *World of Warcraft*, which make no attempt at realism but are nevertheless eminently capable of creating hyperreal worlds for gamers. In any event, there is no reason to doubt that modeling technology will develop to be equal to the task of operational equivalence in the relatively near future.[23]

Hyperreal visuality: Architectural reception

The general public in Western culture has been increasingly immersed in hyperreality for over a century. The recent explosion of Internet-connected mobile devices has extended the reach of hyperreality to our most private moments. The general conflation of art and entertainment is an indication of the generally uncritical attitude of the public towards the media they consume. As a culture, all we expect from our "art" is the thrill of an action movie or the rhythm of a dance track. In terms of architecture, hyperreality thwarts any attempt to communicate ideas in the built environment. Visually, buildings are reduced to appearance.

Architects now design for a public that does not look at the built environment as bearing meaning, as in the traditional sense of the term. As an example, consider a typical contemporary mass-built, single-family home (Fig. 10.1). The interior is usually planned to meet the functional demands of the daily activities of a certain type of family. The exterior has fewer functional determinants, and it is here that simulation becomes most evident.[24] Its design is based on projecting an image and evoking an immediate, unreflective response on the part of the viewer. Apart from meeting a few functional demands, making this impression is the design's only intention. Its referents are vague at best. It cannot refer to the values of its owners – how could it when thousands own a nearly identical house? Very few make real historical references and most of these are hopelessly out of context, robbing the reference of its meaning. Subject to cost constraints, its materials are chosen purely for visual effect; otherwise, how could people accept stone veneer that continues barely a foot back from the front façade? We are in Disneyland here, without the markers that set Disneyland apart as a fantasy world.[25]

Further evidence for the reception of the built environment as hyperreality is the ubiquity of superficial architectural "environments", the nearly universal preference

Figure 10.1 A representative contemporary tract house. Brendel. Public Domain

for pseudo-historical motifs in newly designed public places, the embrace of explicitly simulated environments for everyday activities, and so on. This is being extended and intensified by simulation technologies that are increasingly interposed between people and the built environment. These are both incorporated into the environment itself and carried or worn by people.[26] The former category includes architectural-scale digital displays that transform static walls into animated surfaces and buildings into gigantic signs (Fig. 10.2).

On a personal scale, GPS-enabled smartphones and related devices profoundly alter the individual's experience of the built environment. Through these devices, individuals craft their own experience of which the physical built environment is only one component and a background one at that. As their bodies move through one space, they are likely to be engaged visually and aurally with people and activities taking place elsewhere or nowhere: text messages, phone calls, social media, websites, photographs, videos, personal and multi-player games and so on. If architecture concerns the shaping of spatial experience, these technologies must be regarded as having architectural effects. The melding of such disparate types of experience into a coherent whole can only take place under the sign of simulation. Electronically and physically generated experiences are compatible because their disparate origins are of no consequence under the dominion of hyperreality. As this technology develops to include virtual and augmented reality (Fig. 10.3), the role of media in creating architectural experience will become even greater.[27]

Figure 10.2 RTKL Architects, *LA Live*, Los Angeles, CA. (2010). RTKL.com. David Whitcomb

Figure 10.3 Orators Rostrums and Propaganda Stands, 2012. John Craig Freeman

Possible paths for architecture

Hyperreality cuts the ground out from under any attempt to find or express meaning (in the sense of external reference) in the built environment, or elsewhere in the culture, for that matter. This is a cultural given which architects must face. Every architect must now make a personal choice between two broad alternatives. One is to align their practice with the culture at large: to pursue enhanced performance and create possibly fascinating, if ultimately shallow, experiences. The other is to search for new ways to endow architecture with some kind of significance beyond these limited ends.

There are exciting possibilities for the latter. The same computational tools that facilitate simulation can, if used wisely, allow architects to expand their field of operations and reconfigure the production of the built environment. Subverting the tools to make them serve ends other than simulation requires a sufficiently deep knowledge of them to understand both the constraints they place on practice and the untapped possibilities they contain to liberate it. It also involves rethinking the relationship between architecture and the public. Architecture, like all art, has a responsibility to challenge its public. This entails a further responsibility to create buildings that make it rewarding for the public to accept the challenge. There are at least three general approaches to this.

Preserving representation

Architecture may be able to open cracks in the seamless simulation that envelopes our culture. Architects, by virtue of the persistence of representation in their discipline, are well equipped to recognize the phenomenon of hyperreality and to understand its

fundamental sterility. To cultivate this awareness, architects need to find ways to neu-
tralize the operational tendencies of their tools. One way to do this is to use them in
conjunction with other media. Placing computer models alongside drawings, analytical
diagrams, physical models, material samples, photographs of precedents and context
and so on causes them to assume the aspect of being one way among many to understand
a project, i.e., they become representations. To this end, architects should also display
computer models abstractly and refuse to make realistic renderings, or at least postpone
making them until the design is well resolved by other means. Some clients who expect
to see realistic images of their projects from its early stages will doubtless resist this.
However, explaining to clients how such images deprive buildings of significance will be
an important means of expanding awareness of hyperreality. Architecture schools should
purposefully cultivate this attitude among their students, thus recognizing that they have
many incoming students who experience the world as simulation. Putting drawing at the
center of the curriculum would be a vital part of this program.

Asserting the role of the body in the experience of design and building exposes simulation and returns the design process to one of representation. A traditional design
process based on drawings and physical models, as well as valuing craftsmanship in
production, seems the best way to do this.[28] The computer would become essentially
a support and documentation tool and facilitate the production of models and full-
scale mock-ups.

Fortunately, architects are not alone in facing the challenge of preserving representation in a world dominated by simulation. There many centers of resistance to simulation in our society. Social and political movements that are critical of extant power
structures and ideologies seek to discredit operationally based narratives and reveal
other meanings in events. A prime example is environmentalism, which is engaged in
a constant struggle to assert nature as a reality that exists beyond the view of it as raw
material for human use.[29]

Another center of resistance is the culture of representation itself, preserved in academia, criticism, artistic practices and a segment of the public. The products of this
culture are far less widely appreciated than more popular forms and are often viewed
as elitist. Nevertheless, it remains very strong, having the weight of the Western tradition behind it and a committed group of devotees who wield influence out of proportion to their number. As a part of this culture, architecture's mission is to expose the
workings of simulation and to provoke the public to step out of its customary mode
of experience and take a fresh look at the world.

Embracing experience

Hyperreal experience invokes no ulterior reality. But within this condition, there are
greater and lesser degrees of nuance and complexity in experience. Prior to any sym-
bolism or signification, architecture affects us by shaping our visual and spatial experience. There is a subject who sees, hears, touches, tastes and smells, who experiences
qualities in perception. Gastronomy has a great deal to teach architecture in this
regard. No one thinks a taste represents anything, yet the experience can be nuanced,
unique and suffused with (non-discursive) memory. Architecture has always played on
this register of experience. Logocentrism has suppressed its celebration for centuries.
Paradoxically, simulation may restore it to us.

Eliciting emotional response has been the goal of many architects. It was prominent
in the thinking of Enlightenment architects such as Étienne-Louis Boullée.[30] It was also

Figure 10.4 Studio Roosegaard, *Lotus 7.0* (2010). Studio Roosegaard

the essence of Le Corbusier's distinction between architecture and the "engineer's esthetic".[31] Such thinking, however, was based on a belief that emotional responses to form are universal, so by mastering composition, architects could anticipate the response of the viewer to some extent. We now recognize social and cultural influences on perception and therefore tend to reject such universal notions. This does not prevent contemporary architects from aiming for emotional responses to their work, but it does make guiding principles hard to come by.

Technology is opening up new types of spatial experience. Interactive, responsive and adaptive environments (Fig. 10.4) create dynamic relationships between people and their built environments that have only begun to be considered. So far, these environments have been limited to the scope of installations. They have yet to be explored as complete architecture.

Exploring the problematics of simulation

There are several problematics inherent in simulation that architecture could productively explore. One is the emerging hybrid nature of the subject, both of experience and design. On the experiential side, the electronic prostheses we are increasingly prone to wear offer new avenues by which architects can affect their users' spatial experience. Through augmented reality technology, architectural "interventions" in real environments are becoming possible by designing the information provided by wearable devices. GPS and apps such as Google Maps and Yelp! subtly affect our perception of urban environments by transforming arbitrary locations into landmarks that rival Kevin Lynch's classic urban spatial structures that create our mental maps of a city.[32] While augmented reality has been explored as a medium for visualizing designs,[33] approaching such interventions as design opportunities in themselves is virgin territory for exploration by architects.

On the design side, the complex, flexible, pseudo-intelligent tool of the computer has begun to transform architects' thinking. The hypothesis is that this collaboration produces a new, hybrid designing subject.[34] Computational design processes that combine the radically different capabilities of computer and human produce new species

Figure 10.5 Minimal complexity (2012). Vlad Tenu

of forms whose cognitive bases are unfamiliar. We stand to learn a great deal about human cognition through the architectural investigation of this condition.

Yet another problematic lies in the relationship between computational geometry and human perception. This "fourth geometry"[35] originates, not in our experience, but in the nature of computation applied to geometric operations. One example is the creation of emergent forms through algorithmic design (Fig. 10.5). What is interesting is that this approach grounds architectural form not in objects but in relationships and processes. An important question is whether and how the public will learn to interpret such products as something other than personal expressions. The fact that we find visual order in its products is remarkable if one accepts traditional explanations of the origins of geometric form as generalizations of human experience. The human mind can evidently recognize forms that bear little relationship to previous experience. In that it involves our perception of spatial order, this is a question that concerns architecture directly.

Other ways of thinking architecture in the age of simulation are no doubt possible. The only possibility that architects should not consider is allowing themselves to be borne along by the current of simulation. That is the straight road to the extinction of the discipline. There is more to life than immediate experience and more to meaning than operation.

Hyperreality, technology and performative vision

Baudrillard locates the origins of hyperreality in capitalism, thereby implying that it is an inevitable condition in Western society as presently constituted.[36] On the other hand, it was presumably to empower strategies of resistance to hyperreality that Baudrillard undertook its analysis. In mounting such resistance, architects and others should keep clearly in mind the intimate relationship between hyperreality and modern technology. As pointed out earlier, modern technology's refusal of all values save performance is of a piece with the operationalism at the heart of hyperreality. While the replacement (or "deterrence" in Baudrillard's terms) of meaning by operation has many consequences for architecture,[37] the appropriation of vision by technology is particularly important, since architecture will remain grounded in visual artifacts for the foreseeable future.[38]

Technologically appropriated vision, which may be called performative vision, shares the operationalism that is the hallmark of all technological processes. Performative vision deters questions of the significance of visual experience by equating it with its production by physical, biochemical and neurological processes. It thus excludes subsequent cognitive and imaginative operations that create meaning in the traditional sense. One important consequence of this is that the origin of a visual experience is no longer a factor in assessing its value. In particular, artificially and naturally generated experiences are not valued differently. This does not mean that people are unaware of the differences, but rather that they become able to integrate disparate visual experiences into a single visual field. This not only facilitates the efficacy of visual simulations directly, but further merges simulation and real-world experience by devaluing the difference between the two.

Resisting hyperreality is thus a broader project than bracketing visual simulation off from real visual experience. Performative vision is a direct consequence of modern technology. Opposing modern technology *in toto* is not a viable strategy. We in Western societies are committed to technology as the basis of our way of life and the engine of our prosperity. What is called for are strategies for preserving values in the face of technology that only recognizes the "value" of performance.

Notes

1 David Ross Scheer, *The Death of Drawing: Architecture in the Age of Simulation* (New York: Routledge Taylor & Francis Group, 2014).
2 Jean Baudrillard, *Simulacres et Simulation* (Paris: Galilée, 1981. Published in English as *Simulations*, New York: Semiotext(e), 1983).
3 An exception to this is scientific simulations which are validated by comparing their results with observations of the external universe. By submitting to this test, they acknowledge a reality outside of themselves and are therefore capable of producing genuinely new knowledge. However, even scientists can forget the limitations of their simulations and substitute a simulation for reality. See Sherry Turkle, *Simulation and Its Discontents* (Cambridge: The MIT Press, 2009).
4 Michel Foucault, *The Order of Things: An Archaeology of the Human Sciences* (New York: Random House, 1970), 328.
5 Brian Arthur, *The Nature of Technology* (New York: The Free Press, 2009).
6 Baudrillard writes, "Whereas representation attempts to absorb simulation as false representation, simulation envelops the whole edifice of representation itself as a simulacrum." Baudrillard, *Simulations*, 2.
7 "The real . . . is nothing more than operational." Baudrillard, *Simulations*, 3.
8 In Vonnegut's novel, ice-nine is a fictitious form of water that instantly converts any other form of water into ice-nine on contact. Kurt Vonnegut, *Cat's Cradle* (New York: Holt, Rinehart & Winston, 1963).
9 "Panorama. The public are most respectfully informed, that the subject at present of the Panorama, is a view, at one glance, of the cities of London and Westminster; comprehending the three bridges represented in one painting, containing 1479 square feet, which

appears as large, and in every respect the same as reality. *The observers of this picture being, by painting only, so deceived, as to suppose themselves on the Albion Mill,* from whence the view was taken." (italics added). Text of an announcement of an 18th century London panorama. Source: Panorama (Leicester Square, London, England.) London: James Adlard, 1795(?).

10 "For the first time in the process of pictorial reproduction, photography freed the hand of the most important artistic functions which henceforth devolved only upon the eye looking into a lens. Since the eye perceives more swiftly than the hand can draw, the process of pictorial representation was accelerated so enormously that it could keep pace with speech." Walter Benjamin, "The Work of Art in the Age of Mechanical Reproduction," in *Illuminations* (New York: Schocken Books, 1969), 219.

11 "Although there is a sense in which the camera does indeed capture reality, not just interpret it, photographs are as much an interpretation of the world as paintings and drawings are. Those occasions when the taking of photographs is relatively undiscriminating, promiscuous, or self-effacing do not lessen the didacticism of the whole enterprise. This very passivity – and ubiquity – of the photographic record is photography's "message," its aggression." Susan Sontag, "In Plato's Cave," in *On Photography* (New York: Picador, 1977), 7.

12 In 1954, Marshall McLuhan identified TV as a "cool" medium, i.e., one requiring significant imaginative participation by the viewer. This was based on the low-information content (low resolution and refresh rates) of the technology at the time. McLuhan would probably say that modern TV is a "hot" medium having gained appreciably in information content to closely resemble film. Marshall McLuhan, "Media Hot and Cold," in *Understanding Media: The Extensions of Man* (Berkeley: Gingko Press, 2013), 22–32.

13 The debate about the participation demanded by more recent media such as YouTube and Facebook continues. See, for example, Michael Wesch, "The World Wide Web as a 'Cold' Medium and the 'Practice' of YouTube", *Digital Ethnography Blog*, http://mediatedcultures.net/the-youtube-project/the-world-wide-web-as-a-cold-medium-and-the-practice-of-youtube/comment-page-1/ Accessed January 14, 2016.

14 The notion of virtual reality in journalism has been accompanied since its inception by questions about its journalistic implications, especially the apparent elimination of an explicit point of view, allowing the viewer to forget that what one is seeing is chosen just as carefully as a photograph or video. See, for example, Margaret Sullivan, "The Tricky Terrain of Virtual Reality," *The New York Times*, November 14, 2015, http://www.nytimes.com/2015/11/15/public-editor/new-york-times-virtual-reality-margaret-sullivan-public-editor.html.

15 There is a similar series of audio technologies: stereo, Dolby™, THX™, etc.

16 Architectural design artifacts can also be viewed as stand-alone objects and even works of art, but this is secondary to their referential function.

17 cf. Beatrice Colomina, ed., *Sexuality and Space* (New York: Princeton Architectural Press, 1992).

18 Merleau-Ponty describes this as "direct and primitive contact with the world." Maurice Merleau-Ponty, *The Phenomenology of Perception* (London: Routledge & Kegan Paul, 1962). Translation of *La Phénomenologie de la Perception* by Colin Smith. p. vii.

19 Branko Mitrovic, *Visuality for Architects: Architectural Creativity and Modern Theories of Perception and Imagination* (Charlottesville: The University of Virginia Press, 2013), 80–81.

20 Baudrillard referred to this aspect of hyperreality as "the precession of simulacra". Baudrillard, *Simulations*, 2.

21 This is a matter of technological development to some extent – a good virtual reality environment would make the difference between a model and a building less obvious. Nevertheless, the "brick test" (does a falling brick crush your skull?) will permit differentiation between simulation and reality for the foreseeable future.

22 Hannah Arendt, *The Life of the Mind* (New York: Harcourt, Inc., 1978).

23 Charles Eastman and his co-authors anticipate that building information modeling will eventually replace drawing altogether in both design and construction. Charles Eastman, Paul Teicholz, Rafael Sacks and Kathleen Liston, *BIM Handbook: A Guide to Building*

Information Modeling for Owners, Managers, Designers, Engineers and Contractors (Hoboken: John Wiley & Sons, Inc., 2011), 384.

24 Not that the interiors are any less instances of simulation. "Typical patterns of the daily activities" is a simulacrum that displaces actual life as lived by a fictitious construct that actually describes no one's life. Likewise, "a certain type of family" is a fiction that displaces real families. The real people who eventually inhabit these houses and the real lives they lead are cast into the mold of these simulacra. Attempts to reclaim the house for real life (e.g. decorating and customizing) are usually futile. These simulacra are deeply embedded in its basic planning, which is difficult and expensive to alter. Families generally have no choice but to live in this simulation.

25 These markers hide the fact that our "real" world is no more real than Disneyland. As Baudrillard writes, "Disneyland is presented as imaginary in order to make us believe that the rest is real . . ." Baudrillard, *Simulations*, 25.

26 The day when these too are embedded is not far off.

27 Google Glass™ was an early technology that superimposed virtual information on the physical environment. Its potential was barely scratched before it was withdrawn from the market. Its failure, however, should not be taken as a failure of the idea behind it.

28 Juhani Pallasmaa, *The Thinking Hand: Existential and Embodied Wisdom in Architecture* (Chichester: John Wiley & Sons, 2009).

29 "Everywhere everything is ordered to stand by [by modern technology], to be immediately on hand, indeed to stand there just so that it may on call for a further ordering." Martin Heidegger, "The Question Concerning Technology," in *Martin Heidegger: Basic Writings*, ed. D.F. Kellner (San Francisco: Harper San Francisco, 1977), 322.

30 "Boullée set forth the laws of the beautiful as derived from nature in his 'theory of bodies', a study of the properties of objects laying emphasis on 'their power to stir our senses.'" Jean-Marie Perouse de Montclos, *Etienne-Louis Boullée (1728–1799) Theoretician of Revolutionary Architecture* (New York: George Braziller, 1974), 38.

31 "The Engineer, inspired by the law of economy and governed by mathematical calculation, puts us in accord with universal law. He achieves harmony.
    The Architect, by his arrangement of forms, realizes an order which is the pure creation of the spirit; by forms and shapes he affects our senses to an acute degree, and provokes plastic emotions; by the relationships which he creates he wakes in us profound echoes, he gives us the measure of an order which we feel to be in accordance with that of our world, he determines the various movements of our heart and of our understanding; it is then that we experience the sense of beauty." Le Corbusier, *Towards a New Architecture* (Mineola, NY: Dover Publications Inc., 1986), 1.

32 Kevin Lynch, *The Image of the City* (Cambridge: Technology Press, 1960).

33 Stefan Krakhofer and Martin Kaftan, "Augmented Reality Design Decision Support Engine for the Early Building Design Stage," in *Emerging Experience in Past, Present and Future of Digital Architecture, Proceedings of the 20th International Conference of the Association for Computer-Aided Architectural Design Research in Asia CAADRIA*, ed. Y. Ikeda, C.M. Herr, D. Holzer, S. Kajima and A. Schnabel (Hong Kong: The Association for Computer-Aided Architectural Design Research in Asia (CAADRIA), 2015), 231–240.

34 The computer in this case is not simply automating and accelerating human cognitive tasks, but it is allowing new cognitive processes to take place. This is not to say that computers are capable of cognition. Humans can conceive of cognitive processes that they themselves are not capable of carrying out, but that can be carried out for them by a computer. Cognition is always on the part of the human, first in conceiving the process and then in interpreting its results. The mechanics of the process are carried out by a computer.

35 Robin Evans describes three geometries in his classic book *The Projective Cast*. Computational geometry is considered here as a fourth. See Robin Evans, *The Projective Cast* (Cambridge, MA: The MIT Press, 1995).

36 "Hyperreality and simulation are deterrents of every principle and of every objective. . .it was capital which was the first to feed throughout its history on the destruction of every referential, every human goal, which shattered every ideal distinction between true and false, good and evil, in order to establish a radical law of equivalence and exchange. . ." Baudrillard, *Simulations*, 43.

37 For a comprehensive survey of the effects of simulation on architecture, see Scheer, *The Death of Drawing*.
38 This is obviously true of the displays now used in design interfaces. Immersive (virtual reality) design interfaces are being developed but remain overwhelmingly visual. Experiments are being made to incorporate a tactile (haptic) dimension in design interfaces, but these are limited to involving the hands. Full-body haptic interfaces remain a distant possibility. For the state of current research in haptic interfaces, see, for example, S.B. Schorr, Z.F. Quek, I. Nisky and W.R. Provancher, "Tactor-Induced Skin Stretch as a Sensory Substitution Method in Teleoperated Palpation," *IEEE Transactions on Human-Machine Systems* vol. 45, no. 6 (December, 2015): 714–726.

Bibliography

Arendt, Hannah. *The Life of the Mind*. New York: Harcourt, Inc., 1978.

Arthur, Brian. *The Nature of Technology*. New York: Free Press, 2009.

Baudrillard, Jean. *Simulations*. Translated by Paul Foss, Paul Patton and Philip Beitchman. New York: Semiotext(e), 1983.

Benjamin, Walter. "The Work of Art in the Age of Mechanical Reproduction." In *Illuminations*, edited by Hannah Arendt, 217–251. New York: Schocken Books, 1969.

Colomina, Beatrice, ed. *Sexuality and Space*. New York: Princeton Architectural Press, 1992.

Eastman, Charles, Paul Teicholz, Rafael Sacks and Kathleen Liston. *BIM Handbook: A Guide to Building Information Modeling for Owners, Managers, Designers, Engineers and Contractors*. 2nd. Hoboken: John Wiley & Sons, Inc., 2011.

Ellul, Jacques. *The Technological Society*. New York: Vintage Books, 1964.

Evans, Robin. *The Projective Cast: Architecture and Its Three Geometries*. Cambridge: The MIT Press, 1995.

Foucault, Michel. *The Order of Things: An Archaeology of the Human Sciences*. New York: Random House, 1970.

Heidegger, Martin. "Building Dwelling Thinking." In *Martin Heidegger: Basic Writings*, edited by David Farrell Krell, 343–363. San Francisco: Harper San Francisco, 1977.

———. "The Question Concerning Technology." In *Martin Heidegger: Basic Writings*, edited by David Farrell Krell, 307–342. San Francisco: Harper San Francisco, 1977.

Krakhofer, Stefan and Martin Kaftan. "Augmented Reality Design Decision Support Engine for the Early Building Design Stage." In *Emerging Experience in Past, Present and Future of Digital Architecture, Proceedings of the 20th International Conference of the Association for Computer-Aided Architectural Design Research in Asia CAADRIA*, edited by Y. Ikeda, C.M. Herr, D. Holzer, S. Kajima and A. Schnabel, 231–240. Hong Kong: The Association for Computer-Aided Architectural Design Research in Asia (CAADRIA), 2015.

Le Corbusier. *Towards a New Architecture*. Mineola, NY: Dover Publications, Inc., 1986.

Lynch, Kevin. *The Image of the City*. Cambridge: Technology Press, 1960.

McLuhan, Marshall. *Understanding Media: The Extensions of Man*. Berkeley: Gingko Press, 2013.

Merleau-Ponty, Maurice. *The Phenomenology of Perception*. Translated by Colin Smith. London: Routledge & Kegan Paul, 1962.

Mitrovic, Branko. *Visuality for Architects: Creativity and Modern Theories of Perception and Imagination*. Charlottesville: The University of Virginia Press, 2013.

Pallasmaa, Juhani. *The Thinking Hand: Essential and Embodied Wisdom in Architecture*. Chichester: John Wiley & Sons Ltd., 2009.

Panorama. *Panorama*. London: James Adlard, 1795(?).

Perouse de Montclos, Jean-Marie. *Etienne-Louis Boullee: Theoretician of Revolutionary Architecture*. New York: George Braziller, 1974.

Peter Eisenman, Jacques Derrida. *Chora[L] Works*. New York: Monacelli Press, 1997.

Scheer, David Ross. *Architecture as an Ethics of Technology*. September 15, 2014. http://deathofdrawing.com/architecture-as-an-ethics-of-technology (accessed January 5, 2016).

———. *The Death of Drawing: Architecture in the Age of Simulation*. New York: Routledge Taylor & Francis Group, 2014.

Schorr, S.B., Z.F. Quek, I. Nisky and W.R. Provancher, "Tactor-Induced Skin Stretch as a Sensory Substitution Method in Teleoperated Palpation." *IEEE Transactions on Human-Machine Interactions* (IEEE) 45, no. 6 (December 2015): 714–726.

Sontag, Susan. *On Photography*. New York: Picador, 1977.

Sullivan, Margaret. "The Tricky Terrain of Virtual Reality." *The New York Times*, November 14, 2015, http://www.nytimes.com/2015/11/15/public-editor/new-york-times-virtual-reality-margaret-sullivan-public-editor.html?_r=0.

Turkle, Sherry. *Simulation and Its Discontents*. Cambridge, MA: The MIT Press, 2009.

Vonnegut, Kurt. *Cat's Cradle*. New York: Holt, Rinehart and Winston, 1963.

Wesch, Michael. "The World Wide Web as a 'Cold' Medium and the 'Practice' of YouTube." *Digital Ethnography Blog*, http://mediatedcultures.net/the-youtube-project/the-world-wide-web-as-a-cold-medium-and-the-practice-of-youtube/comment-page-1/ (accessed January 14, 2016).

11 Rebooting spaceship earth
Astrospatial visions for architecture and urban design

Davina Jackson

Introducing another space era

"Spaceship Earth" is back on the agenda with futuristic architects and environmental planners. Popularised by Richard Buckminster Fuller and other modern science pundits during America's 1960s space race against Russia, this term[1] remains the most evocative of several concepts which promote the accelerating ambition to manage holistically our planet's environmental systems. One ancient seer was Claudius Ptolemy, whose second-century Latin treatise, *Geographica*, clarified most of the basic principles of mapping the globe by observing the night sky. He wrote that the essence of cartography is to "show the known world as one and continuous" and that "the first thing [to understand] is Earth's shape, size and position with respect to its surroundings . . . the celestial sphere".[2] Now his astronomical strategies are rivalled by a reverse perspective: to ubiquitously monitor Earth *from* space (Fig. 11.1).

In this century, the Spaceship Earth dream is being facilitated by telecomputation tools originally devised to fly aeroplanes, rockets and satellites. Pulsing the scenes flickering across our myriad screens are the semiconductor and sensor-enabled infrastructures of "massive parallelism"[3] connecting non-visual data across globally distributed grids of processors, portals and storage banks. As predicted by Al Gore in his 1992 proposal for a "Digital Earth" global climate model, parallelism seems to be the only systems architecture, and conceptual metaphor, that could "cope with the enormous volume of data that will be routinely beamed down from orbit".

How will all these bits of information help architects to envisage structures made of atoms? This question, published in 1995 by William J. Mitchell to extrapolate the urban development implications of common access to the Internet,[4] still highlights the crucial paradox and paradigm for professionals dealing with virtual architecture:

> The network is the urban site before us, an invitation to design and construct the City of Bits (capital of the twenty-first century). . . . But this new settlement will turn classical categories inside out and will reconstruct the discourse in which architects have engaged from classical times until now. . . . How shall we shape it?[5]

Today, astrospatial (developed for space exploration) and geomatics technologies are propelling the hybrid domain of Earth observations (EO), which underpins fundamental reforms of geography, surveying and environmental planning.[6] Emerging systems of visualising today's torrents of location-specified data (with Landsat-style 2D satellite images and/or *a posteriori* processes such as photogrammetry or 3D lidar scans converted to point clouds) also require major innovations to help merge different ways of

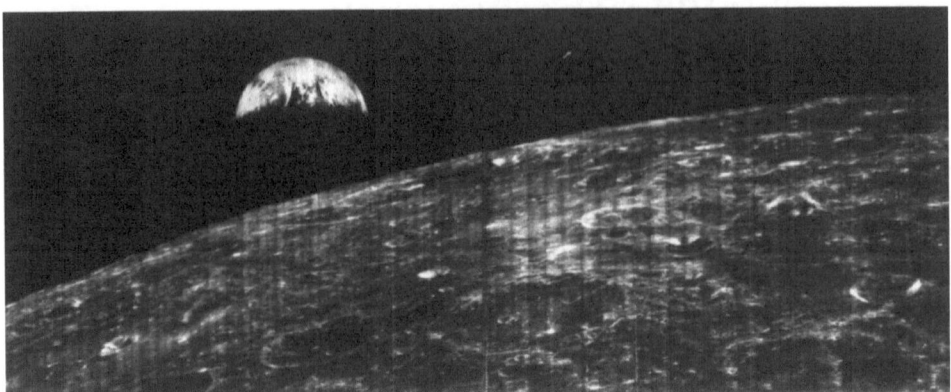

Figure 11.1 Image of the world from the US Army's V-2 #13 missile, October 24, 1946. White Sands Missile Range/Applied Physics Laboratory

modelling natural and constructed environments. Third-millennium computer simulations, comprehensively explained by Stephen Wolfram in *A New Kind of Science* (2002), are being underpinned increasingly by his "Principle of Computational Equivalency" between complex natural processes and their correct mathematical models (which may be generated by surprisingly simple cellular automata programs).[7] Wolfram's concepts are accelerating various compatible practices across many long-estranged science disciplines[8] and are unlocking (for advanced architects and other building professionals) digital simulation scenarios that go far beyond the current capacities of CAD-CAM, BIM, GNSS (GPS), GIS and sg.[9] Textbooks explaining behavioural modelling for a "new science of cities" were published by Michael Batty (2005, 2013)[10] after his UCL colleague, Andrew Hudson-Smith, wrote a comprehensive thesis (2003) on virtual visualisation technologies relevant to online urban planning.[11] Spatial techniques to represent urban flows are exemplified by mobile phone data videos of pedestrians and buses (MIT SENSEable City Lab, 2006, and many later examples) and are expanding architecture's core premise to envisage (static) building stocks.[12]

Two recent *Spaceship Earth* exhibitions at Customs House in Sydney (2014) and Ars Electronica in Linz (2015)[13] intrigued audiences with high-resolution images of earthly environments captured using specialised cameras, radar scanners and other electromagnetic wave-detection equipment aboard EO satellites operated by the European Space Agency (ESA), NASA and the main commercial constellation operator, DigitalGlobe.[14] Since 1995, these organisations, and affiliates, have been supplying satellite imagery to Google, Apple, Microsoft, Esri and other public providers of digital cartography and sat-nav (mainly the United States GPS) services, but the precision of co-ordinates data and resolution of map imagery often have been too coarse for use in detailed architectural representations.[15] Resolving diverse gaps between knowledge, ideas and practices in different disciplines is challenging many scientists, engineers and strategists across the space, computer systems, defence, logistics, infrastructure, property, governance, design, construction and digital media industries.

Google Earth's commercial launch in 2005 catalysed an alternative schema for digital architecture – from how maps, traditionally read as static 2D drawings, inform designs for 3D buildings, to a need for online architectural models to integrate with streaming simulations of urban (including natural) flows and – in the ultimate vision – modelling of the continuous, complex, systems of our entire planet. Both approaches may be valuable, in different ways, during project design and testing, but the notion of assimilating architectural models within an immense Earth systems simulation basically contradicts the discipline's Vitruvian lineage – where stasis and permanence ("firmitas")[16] have been key priorities to ensure monumentality.[17]

This crucial conceptual shift, enabled by new (information) architectures of sight, has stimulated some revolutionary (physical) sights of architecture. Some can be experienced synaesthetically (with multiple senses) rather than only optically (by arousing awe through the eyes). Two early twenty-first-century triumphs of multisensory aesthetics were Diller Scofidio and Renfro's mist-shrouded Blur pavilion on Lake Neuchâtel, Switzerland (Swiss National Expo, 2002)[18] and Carlo Ratti Associati's torrent-curtained Digital Water Pavilion at Zaragoza, Spain (International Expo, 2008).[19] Both substituted solid walls for mantles of moisture and stimulated observers with different sensory experiences of space rather than optically sublime forms. As well as visually arousing, yet ephemeral, video and light effects, these venues included sensor systems configured to surprise and engage visitors with dynamic sound, smell, touch-responsive and skin-tingling effects.

Modelling global earth systems

More than any physical structure, Google Earth (GE) was the artifice which explicitly highlighted massive urban implications from our escalating "space economy"; a phenomenon surveyed by the Organisation for Economic Co-operation and Development (OECD) since 2007.[20] While GE is only one of various online virtual globes,[21] it was the first to show why it has been intelligent for earthlings to explore outer space. It demonstrated that we must rely on ubiquitous surveillance from orbiting vehicles to realistically comprehend our planet's conditions: visible and invisible, spatially from core to stratosphere and horologically from genesis to oblivion.[22] Inspired by the legendary Blue Marble photograph from NASA's Apollo 17 astronauts (1972)[23] and *Powers of Ten* short film by Charles and Ray Eames (1977),[24] then boosted by NASA's Mission to Planet Earth program (since 1988),[25] GE astonished viewers with its brain-spinning, point-to-point transits and veristic depictions of terrestrial scenes. Although its pixel-deep photogrammetry interface must be reinvented to effectively illustrate vast flows of diverse data in conceptually infinite dimensions and combinations, GE seems to be the most profound science-art contribution to visualising our world since Eratosthenes of Cyrene calculated the circumference of Earth, drew the first gridded maps of Alexander's empire and clarified other fundamentals of geography.[26]

Fundamental discoveries in environmental science usually precede (and provide contexts for) conceptual advances in architecture. For example, Eratosthenes' *Geographica* trilogy (which influenced Ptolemy) is presumed to have been written during 240–220 BC, which was approximately two centuries before Vitruvius wrote his canon of ten architectural manuals.[27] Surveying and maps always have been essential to evolving the world views of humans and societies and (more recently) to defining the contexts for designing, engineering and regulating land developments. Many practitioners now use GE's satellite imagery as a starting point for site planning: from showing clients how their buildings may be overshadowed or affect neighbours, to representing different aspects of site context for stakeholder reviews of precinct proposals.

A key advantage of GE and other virtual globes is their potential to support various kinds of 2D and 3D visual representations of different kinds of data (mash-ups). Yet there is a problem of unpredictable misregistration from errors in patching together the base imagery and incorrectly locating landmarks at precisely correct (or at least consistent) co-ordinates. Michael Goodchild et al. (the International Society for Digital Earth's Vision 2020 group) reported in a 2012 *PNAS* article that (in GE)

> distances, for example between Los Angeles and New York, can be reported in hundreths of centimeters, an absurd prediction given the lack of an accepted definition of the distance between two extended objects. . . . The actual amount of misregistration of base imagery varies over space and over time, depending on the availability of small, recognisable features with known locations that can be used as reference points, on the spatial resolution of the imagery itself, and on many other factors. . . . Over Santa Barbara, CA, misregistration has been as much as 40 m at times, relative to high-quality GPS measurements. What is important, of course, is not whether misregistration occurs, as it must, but whether it is substantial enough to affect a given application. Moreover, positional uncertainty is only one of the many dimensions of uncertainty that characterise geographic data.[28]

Reinforcing the third-millennium phenomenon of GE and other virtual globes, scientists and technically literate policy strategists have been updating Fuller's Spaceship Earth and Gore's Digital Earth scenarios as an international environmental monitoring and modelling scheme called the Global Earth Observation System of Systems (GEOSS). Proposed at the World Summit on Sustainable Development in Johannesburg in 2002, progressed at two following Earth Observation Summits in Washington, DC (2003), and Tokyo (2004), and launched in Brussels (2005, the same year as Google Earth), this ten-year intergovernmental program recently was approved to continue to 2025 (at least). GEOSS is co-ordinated by the Group on Earth Observations (GEO),[29] a secretariat within the World Meteorological Building on the United Nations campus in Geneva. Directed by American geographer Barbara Ryan (who drove the 1990s open data campaign to freely release NASA's once-expensive Landsat imagery), GEO includes one hundred member nations and ninety-three global science and non-government, non-commercial organisations.

Intriguingly, the utopian GEOSS environmental surveillance project is not yet widely known or discussed among built environment academics and has been ignored by the general media. News channels instead perpetuate political arguments about emissions reduction sanctions and carbon trading schemes, all responding to the Intergovernmental Panel on Climate Change (IPCC) assessment reports, released every five to seven years,[30] and the United Nations Framework Convention on Climate Change (FCCC) "conferences of the parties" (COPs), which are held in a different city each December to update the Kyoto Protocol.[31]

GEO (another intergovernmental group that is co-operating with the United Nations on climate-alleviation challenges) is developing a global online exchange for supplying and accessing public environmental datasets, which often must be visualised (frequently with dynamic map platforms) to be comprehensible to audiences lacking statistical analysis skills. GEO-aligned agencies are collaborating to help strengthen and openly disseminate distributed banks of datasets that are needed to enable consistent future analysis of more than fifty "essential climate variables" (ECVs) that have been defined to underpin the Global Climate Observing System (GCOS).[32] British space policy strategist Stephen Briggs (advising the European Space Agency) has said that most of these climate variables – atmospheric, oceanic and terrestrial – can only, or mainly, be surveyed from satellites, and consistently for the multi-decade timescale needed to reveal significant behaviour patterns.[33]

How do satellite Earth observations work? Here are the most salient points. EO satellites (usually unmanned but including the astronaut-occupied International Space Station) mainly are operated by government space agencies – NASA, the European Space Agency (ESA), Roscosmos (Russia), JAXA (Japan), CNSA (China), ISRO (India) and KARI (South Korea). The first and largest commercial operator is DigitalGlobe, but several hundred new businesses and universities also now are operating sensor-carrying nanosatellites (including CubeSats, measuring only $10 \times 10 \times 10$ cm) for different experimental and entrepreneurial purposes. On board these solar-powered satellites are monitoring kits that include active and/or passive sensors which measure (and transmit to Earth-based computers) certain frequencies of electromagnetic waves that capture specific kinds of signals about the surveyed zones. Active sensors project radar signals to the target zone and show how the electromagnetic waves are reflected, refracted or scattered by Earth's surface or atmosphere. Passive sensors are infrared or microwave instruments that detect and reveal emissions naturally produced by Earth and/or its atmosphere.

Another critical factor in understanding satellites is their height of orbit. Many EO and nano-satellites operate in low Earth orbit (the upper atmosphere). The American global positioning system (GPS) satellites rotate in medium Earth orbit (about 20,200 km above Earth) and many weather-monitoring satellites operate in high Earth orbit (about 36,000 km above Earth's surface). Also critical is the type of orbit: either geo-synchronous (aligned to one point on Earth) or sun-synchronous (aligned to the sun so as to see each part of Earth during the same hours, to avoid inconsistencies caused by differences of light).[34] All of these satellites produce EO data which, in visual formats, especially through map apps on mobile devices, are constantly watched by humans (Fig. 11.2).

Figure 11.2 Central Dubai seen from a Digital Globe satellite. Digital Globe

Online environmental planning

What does global monitoring from space vehicles have to do with architects of terrestrial buildings? It seems logical that government planners in the future will require property developments to be designed and cross-checked against locally relevant environmental datasets and that ECV data and other natural systems information will need to be integrated (more visually) with design modelling of major building projects. EO surveillance, using equipment to measure invisible electromagnetic waves reflected from land, sea and air surfaces, seems especially valuable to help clarify whether sites are suitable, or not, for future human living. One example was the 2004 Arup-planned proposal to build China's first eco-city, named Dongtan, on a swampy island off Shanghai. After substantial international publicity that was later described as "greenwash",[35] viability for this project evaporated after scientists revealed not only that construction would slaughter local wildlife but that Dongtan is among many coastal and island locations that will be submerged by progressively higher tides. Previously, sea-surface heights were not considered often in architectural design, but today's satellite-informed forecasts are becoming more pertinent: not only for planning seaside cities but also in designing and insuring houses for oceanfront, clifftop and flood-prone sites.

Conversely, Earth monitoring may have the potential to highlight large land areas, for example, the deserts of North Africa and Australia, which may become suitable habitats for humans (via global warming or other climate changes, or with substantial engineering). Architects already are imagining fantastic scenarios which, via photorealistic CGI, may convince (unfamiliar) viewers that they have been built. For example, Manal Rachdi (OXO, Paris; Fig. 11.3) has published a 450-m-tall vertical city concept

Figure 11.3 Manal Rachdi's proposed "City Sand Tower". OXO Architectes

proposing solar and geothermal power and rainwater collection to sustain a tower of offices, a hotel, community facilities and six hundred housing units, suggesting this should be sited in Morocco.[36] Swedish architect Magnus Larsson visualised a new Sahara "dunescape", incorporating a 6000-km-long "shelterbelt" of trees and a sand-sculpted desert camp comprising caves for several thousand refugees. With this project for his 2008 AA diploma thesis, Larsson used ambitious bio-chemistry strategies (which Rachel Armstrong has generically termed *Vibrant Architecture*)[37] of sowing *Bacillus pasteurii*, a wetlands bacteria, to transform the sand particles into a structur-ally cohesive, fibrous stone structure with cavities that could be occupied by otherwise homeless humans.[38] Rachdi, Larsson and many other speculative architects seem inspired to draw solutions for the current international governance challenge of find-ing new homelands for large groups of refugees from wars or natural catastrophes. Presumably, today's hyper-realistic architectural renderings are helping to stimulate new ideas for development of non-obvious sites and fresh markets for providers of built environment services.

Monitoring Earth's physical resources requires government agencies to collaborate to establish a global computational infrastructure for managing the gathered informa-tion. The Group on Earth Observations has UN support and funding from thirty countries to globally advocate the GEOSS project, engage stakeholders and co-ordi-nate data systems that could effectively inform environmental management decisions. GEO promotes best practices already achieved by its various member organisations – notably the European Commission's INSPIRE (spatial data infrastructure) initiative, which has established thirty-four categories of data that (by European law since March 2007[39]) must be collected and managed by EU government agencies to help underpin a comprehensive, consistent environmental monitoring and modelling system.[40] INSPIRE is intended to provide more detailed territorial information than the GCOS ECV datasets.[41] Gradually, INSPIRE is influencing many other nations' strategies to develop national spatial data infrastructure systems (NSDIs). Encouraged by the United Nations' Global Geographic Information Management (UN-GGIM) offices in New York and Brussels, NSDIs are expected, eventually, to become robust enough to help reform antiquated methods of planning, designing and regulating land developments.

Local and provincial governments already operate their own SDI systems – all based on the traditions of cadastre: one official government land survey, map and records archive of the real estate information needed by an area's rulers for deciding taxes and controlling property rights. Like cartography and surveying as ancient disciplines, cadastres now are evolving from 2D static maps on paper to nD (theoretically infinite dimensions and domains) of information, most of which must be communicated between semiconductor-enabled devices, often without being visualised for the eyes of humans. Today's massive mutations of cadastre (and census) administration systems are driving today's global debates about smart cities (now led by global technology and systems suppliers fighting to win, roll out and lock in major government contracts) and data cities (exploring how to collect, manage, contribute and use data to solve globally prevalent urban challenges).

Currently, there is a vacuum of coherent recognition about how public open online access to extensive repositories of environmental data might help local planning and development professionals to more effectively serve their constituencies and contrib-ute responsibly to global climate management strategies. Minimising corruption of data – during gathering, storing, analysing, exchanging and disseminating – looms as a colossal challenge.[42] Public transparency strategies are suggested by many promoters of the GEOSS project. Yet some governments now aim to privatise their land and property information agencies – suggesting that data access will become a fee-for-service.

Conceptually, at least, national and local spatial data infrastructure systems seem essential to seriously consolidate today's rhetoric about "evidence-based planning",[43] which is a policy context that increasingly controls twenty-first-century practices of designing architecture. Three international data-agglomeration movements are evolving and all depend on both automated computation and visual representations to make sense of the raw content. First, many advanced governments now have programs to location-tag as much public information as possible, especially census statistics: this thrust may help transform representations of cities from 2D and static mapping to 3D (and conceptually nD) dynamic models. Another new scheme is ISO 37120:2014, the world's first standards code to support comparisons of municipality performance indicators, which was prepared by the Global City Indicators Facility at the University of Toronto and adopted by the International Standards Organization in 2014.[44] An earlier concept, launched by UN-Habitat at its Habitat II conference in Istanbul in 1996 and prototyped by some Middle East and North African cities since, is the GUONet global network of "urban observatories": centres for collating, analysing and publishing (mostly graphically, using 2D maps) statistics recording location-relevant social and environmental conditions.[45]

The urban observatories idea was conceived by American information architect/author Richard Saul Wurman, beginning with his same-scale plasticine models comparing the land contours of fifty different cities (1963).[46] In *Design Quarterly 80: Making the City Observable* (1971), he reviewed the potentials for visual evidence (of urban stocks such as buildings and flows of traffic or natural forces such as wind and water) to inform more accurate development decisions. He proposed two types of clearing houses: "urban observatories" (for monitoring and analysis) and "urban data centres" (for storage and access).[47] These distinct, yet interlinked, operations still seem vital to underpin a globally congruent system for planning and managing future urban developments and would need science-astute professionals, remotely supported by supercomputers, to facilitate valuable uses of the data. Obviously, there is a substantial gulch between this ideal, current supercomputing limitations, resistance to the UN GUONet project so far, and today's *realpolitik* of urban planning around the world. Yet Wurman's prescient strategies remain conceptually essential to evolve new architectures, dependent on data visualisations, to help humanity manage our "spaceship".

> Public information should be made public. Information about our urban environment should be made understandable. Architects, planners and designers should commit themselves to making their ideas immediately comprehensible.[48]

During its first ten years of operation, GEO's list of nine "societal benefit areas" (SBAs, predicted to flow from future users of GEOSS-mediated environmental data) was biased towards big agendas for managing natural systems: water, weather, climate, biodiversity, ecosystems, energy, agriculture, disasters and health. In its new GEOSS strategy plan for 2016–2025, eight SBAs now include two domains – sustainable urban development and transport-infrastructure – which are potentially relevant to building and urbanism professionals.[49]

Earth observation outcomes: Data city systems and geodesign

How will the GEOSS affect architectural practice? And (how) will architecture practitioners contribute to this scientific project? If implemented successfully over the next decade, the GEOSS would provide access (theoretically through one online portal) to many globally distributed banks of the geo-tagged and climate-related information that seems necessary to underpin evidence-informed designs for future places to live. The point of all this data, relevant to architecture, is that architects will be expected to exploit it not only for specific projects but also to continue to reform the profession's methods of design and representation. Today's vital innovations are coming not just from visualisation software suppliers (Autodesk, Esri, Trimble, Hexagon, Bentley and others – noting Google-Alphabet's Sidewalk Labs spin-off), but also from the research departments of major inter-disciplinary professional consultancies (for example, AECOM, Aedas, Fosters, MVRDV,[50] Frank Gehry,[51] Zaha Hadid, Greg Lynn FORM, Heatherwick Studio, Arup, Buro Happold and many engineering firms). Academia's contributions include progressive international research-conference networks (such as Smartgeometry[52] and the five regional CAAD groups[53]) and agenda-setting postgraduate centres at various universities (notably the CASA;[54] Space Syntax[55] and architecture units at UCL's Bartlett faculty; the MIT Media and SENSEable Cities Labs; the ETH-Z Future Cities Labs; the Dutch, German and Austrian TU systems; and schools like the AA in London and IAAC in Barcelona).

Underlying recent debates about the latest long wave of climate change is a shared concern about how (or whether) humanity can avoid massive losses of life and (in some projections) eventual extinction as a species.[56] Scientists promoting integrated Earth systems simulations suggest that "visual computing"[57] of multiple dimensions of information is essential to clarify patterns of activity and insights towards effective solutions. Bob Bishop, chairing a campaign to build an International Centre for Earth Simulations[58] in Geneva (comparable in his proposed scale to the CERN particle physics facility), suggested that "one important advantage for visualisation-based analysis is that computer simulation output can be presented as multiple layers of data for every time-step in a process".[59] One example is the Australian Geoscience Data Cube,[60] launched in 2013 by Geoscience Australia's National Earth Observation Group to save substantial time in analysing time-sequences of NASA Landsat imagery over Australia's vast land mass (which spans forty zones of longitude and latitude). The Data Cube process is to slice the Landsat imagery into tiles covering precisely the same land coordinates, "stack" the tiles as a time-series, and then identify, extract and analyse only the differences of data for each specific zone. While the idea of layering datasets was dramatically demonstrated with Apple's Time Machine method of storing and visualising document files (released in 2007), the idea had not been applied usefully to satellite imagery. Now supported by influential EO agencies,[61] this system seems likely to accelerate the ease of analysing how natural systems affect areas with potential for building developments.

Another promising demonstration of advanced analytics of satellite images of built environments is the European Commission's Global Human Settlement Layer (GHSL) launched in 2013.[62] An EC science team at Ispra, Italy, developed a high-performance computing process named I2Q to automatically query sensor and population data from satellite and aerial images of cities and settlements. Tested on all types of sensors, the I2Q-GHSL system can be used to visualise built surfaces, percentages of built surfaces, average sizes of buildings and the numbers of structures for every image (tile). This introduces a widely applicable automatic processing method to generate globally consistent, optimised mapping of the structural conditions of settlements, thus supporting international responses to crises and the sustainable urban development agendas of the UN and GEO systems (Fig. 11.4).[63]

Advances from scientists towards a sophisticated global system of simulating environments are not seriously accessible or cohesive yet for architects, and most geomatics experts are not equipped to design physical facilities for urban areas: this void seems propitious for entrepreneurs. Also not concerned with designing real cities – but far advanced in visualising fantasy environments and simulating real cities (as in film scenes where famous monuments seem to explode) are CGI studios such as Pixar and Weta, which create awesome environmental movies for the entertainment, exhibition and advertising industries.

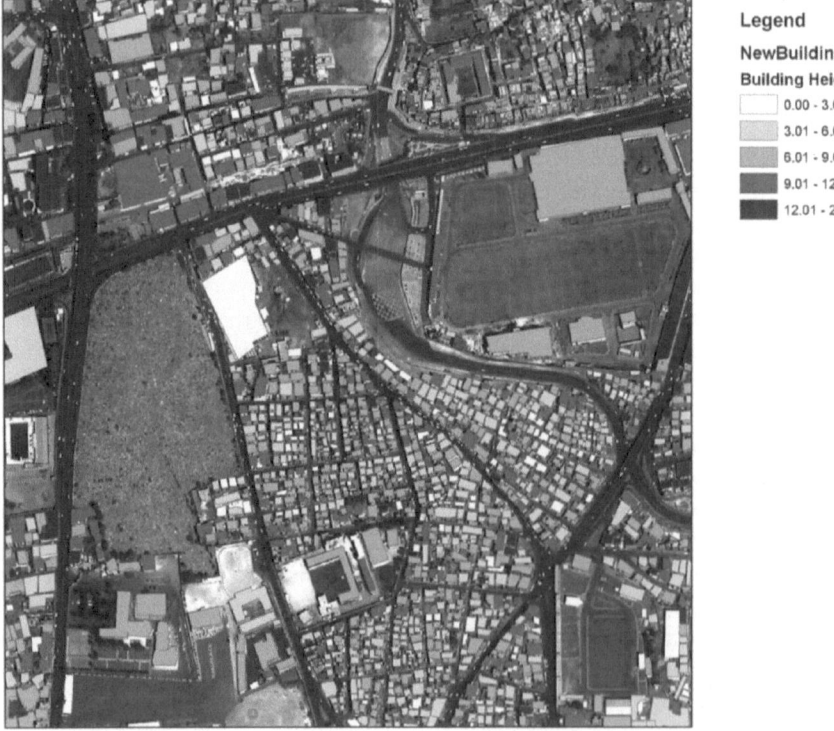

Figure 11.4 Map of Sana'a, Yemen. EC JRC, Ispra/Institute for the Protection and Security of the Citizen

An emerging technique for environmental simulations is procedural modelling, where scenes are algorithmically generated in commercial programs such as CityEngine or Houdini, or using open software such as GML (generative modelling language). Procedural programs, which generate virtual forms through computational processes (operations), inherently can save considerable time in modelling urban areas comprising many buildings, while parametric modelling, where structures are assembled with specific rules and measurements, produces more accuracy in detailing but is less flexible for changing basic design strategies. Parametric modelling (the approach underpinning "smart geometry") is especially valuable for architects designing geometrically irregular buildings, especially those incorporating complex curves. Procedural modelling is proving useful for urban designers wanting to change and compare possible envelopes (forms) of multiple buildings in urban locations, and it is impressive in stakeholder communications about new concepts for urban precincts.

All modelling methods (including non-digital) are encouraged by protagonists of the Geodesign (GD) movement, which Esri has promoted at special user conferences since 2010. Carl Steinitz, the former Harvard landscape professor who authored the first Geodesign manifesto,[64] has said that GD modelling requires both design arts and geographic science skills, in different proportions and using different processes, to help answer the following six questions: 1. How should the context be described (Representation models)? 2. How does the context operate (Process models)? 3. Is the current context working well (Evaluation models)? 4. How might the context be altered (Change models)? 5. What differences might the changes cause (Impact models)? 6. How should the context be changed (Decision models)?[65]

One key to potential convergences in modelling buildings, cities and their natural contexts is the increasingly prevalent technology arena of lidar (laser and radar scanning), where cameras pan mirror-reflected beams of coherent waves across sections of a target area from either tripods on the ground or low-flying aircraft. Lidar scans are stitched together as grainy images called point clouds, with every point of an image precisely representing specific x-y-z co-ordinates in airspace. This method produces precisely geo-tagged simulations of the surfaces of complex structures, such as historic stone monuments with intricately carved details, and can provide far more detailed architectural information than is possible with other imaging methods. Different qualities of survey data are obtained from light aircraft, drones (UAVs), balloons, moving trucks or fixed tripods. Generally, the terrestrial and low-flying scanners obtain higher-quality resolutions than satellite images – but they survey targets only once or infrequently, while satellites now promise constantly updated information to inform modelling over multiple decades.

Also transforming traditional visions of architecture are many designers and artists who exploit cities as after-dark stages for "licht architektur"[66] spectacles, using post-Edison (mostly RGB LED) illumination sources and control systems. Electroluminescent (semiconductor-enabled) urban lighting technologies are equipping the most profound new arts and architecture movement of the early twenty-first century. Breaking out from conventional exhibition containers, such as galleries and museums, today's "smart light cities"[67] savants are transforming buildings, bridges, streets, skylines, fountains, waterways, parks and public spaces with multi-storey video projections; facade-scale pixel screens; gobo (stencil)-filtered pole lights; luminous footpath substances; and radiant, technicolour sculptures with invisible sensors responding to human presences, touches, voices, steps and eye movements. Experimental

Figure 11.5 Paris artist Yann Kersalé's Sea Mirror heliostat of colour-changing LEDs and mirrors, cantilevered from an upper floor of Sydney's Central Park One apartment tower, designed by Jean Nouvel with Peddle Thorp (Fraser Properties-Sekisui House)

protagonists include Yann Kersalé (Fig. 11.5), Hervé Audibert, Hervé Descottes (L'Observatoire International), Rafael Lozano-Hemmer and Daan Roosegaarde – influencing many international designers of architectural lighting and electronic theatrics for public events. Futuristic architecture faculties are improving their resources to educate students about emerging techniques of architectural lighting. Technology-experimental, cross-disciplinary researchers seem to be pioneering novel transmedia – and transarchitecture[68] – genres where citizens may drift between virtual and physical domains of behaviour. Diverse experiments with new ways of experiencing light and visually exploiting data are being encouraged especially by (for example) the Media Architecture Institute (emphasis on urban and architectural light experiments)[69] and the International Society for Presence Research (emphasising scientific advances with virtual reality, augmented reality and robotics).[70]

Optics – the science domain concerned with light and vision – has always catalysed the concepts which philosophically advanced architects interpret as themes for constructing the aesthetics of buildings. During the United Nations' International Year of Light 2015, optics was promoted as the source of most of the electroluminescent (including data-conducting) technologies which are transforming our cities and ways of life. Current theories of quantum electrodynamics (QED), clarified by Richard Feynman in lectures from 1979,[71] interpret all optical and electromagnetic behaviours in terms of dynamic exchanges between electrons (particles of matter) and photons (particles, or what he called 'corpuscles', of light). Feynman's principles are vital for twenty-first-century interactions between real and virtual worlds: they explain many emerging strategies for information transmissions such as li-fi, data modelling, light effects such as holograms, and virtual and augmented reality.

Astrospatial architecture: Design in digital space

What would R. Buckminster Fuller think of today's explosion of post-Edison, semiconductor-controlled electrical systems and their potentials to accelerate his "energetics-synergetics" theories? As well as his geodesic architectural shelters and engineering of vessels and vehicles, Fuller progressed a global logistics vision from his 1928 manifesto, *4D Timelock* (including an axonometric ink sketch of life on Earth, promoting his "Air-Ocean World Town Plan") to his late-1960s book, *An Operating Manual for Spaceship Earth*, and posthumous (1992) volume, *Cosmography*.[72] His recognition of light as a crucial transmitter of computer data is evident in his sophisticated proposal for a "Mini-Earth" exhibit, written for the American Institute of Architects in 1963 – four years before his legendary geodesic pavilion opened at the Montréal Expo '67. The following is an abridged extract:

> The design of a two-hundred-foot diameter Miniature Earth. . . . fabricated of a light metal trussing. Its interior and exterior surfaces could be symmetrically dotted with ten million variable intensity light bulbs and the lights controllably connected up with an electronic computer. . . . Information could be "remembered" by the computer, regarding all the geographical features of the Earth . . . under a great variety of weather conditions. . . . If we use the thirty-five millimeter contact prints of the photographs taken by the aerial surveyors . . . Man on earth . . . would be able to see the whole Earth and at true scale in respect to the works and habitat of man. He could pick out his own home. Thus Mini-Earth becomes a potent symbol of man visible in the universe.[73]

More than half a century after this AIA lecture – and one decade into the Google Earth (GEOSS) era – these word-pictures seem almost quaint. Fuller died in 1983, two years before Feynman published his seminal book of lectures updating 1920s theories of quantum electrodynamics,[74] but he already must have recognised that QED would unlock many novel applications of his "universal architecture" and "world planning" dreams. Today, the technologies of light waves – whether visible or not to humans – are propelling a new global Enlightenment age – including a future-habitats design movement that could be named astrospatial architecture.

The physical frontiers of astrospatial architecture already could be claimed to extend to the moon and Mars, which are targets for increasingly serious research, design and 'analogue' (earthly) prototype tests involving "space architects"[75] who focus on how earthlings might comfortably live in spacecraft or on other planets. Among leaders of this new domain are architectural research teams at Foster and Partners, London, which has published NASA-supported schemes for lunar and Martian building complexes that would be 3D-printed by robots, using local soil. Fosters also has designed Spaceport America, the world's first space travel terminal, in New Mexico.

However, designing physical structures for real localities – whether earthly or otherworldly – is not really the key design or domain distinction of astrospatial architecture. This emerging realm of creativity is mediated entirely via planes of pixels separating human occupants of airspace from cybernetic constructs assembled in what neogeographer Andrew Hudson-Smith termed "digital space."[76] He noted (2003) that "digital space takes many forms, and it is limited only by our imagination."[77] Today, that seems like a useful general axiom to help perceive revolutionary potentials for astrospatial architecture on, and beyond, Spaceship Earth.

Notes

1 "Spaceship Earth," *Wikipedia*, accessed January 11, 2016, https://en.wikipedia.org/wiki/Spaceship_Earth. Buckminster Fuller, *An Operating Manual for Spaceship Earth* (New York: E.P. Dutton, 1968).

2 J. Lennart Berggren and Alexander Jones, *Ptolemy's Geography: An Annotated Translation of the Theoretical Chapters* (Princeton NY and Oxford: Princeton University Press, 2000), 57–59.

3 Al Gore, *Earth in the Balance: Ecology and the Human Spirit* (New York: Houghton Mifflin Harcourt, 1992), 358.

4 William J. Mitchell, *City of Bits: Space, Place and the Infobahn* (Cambridge, MA: The MIT Press, 1995).

5 Mitchell, *City of Bits*, 24.

6 Davina Jackson and Richard Simpson (eds.), *D_City: Digital Earth | Virtual Nations | Data Cities [Global Futures for Environmental Planning]* (Sydney: DCity and Geneva: GEO, 2013). *D_City*, accessed January 11, 2016, http://dcitynetwork.net/manifesto.

7 Stephen Wolfram, *A New Kind of Science* (Champaign, IL: Wolfram Media, 2002).

8 Wolfram, *A New Kind of Science*, 7–16.

9 Abbreviation meanings: CAD-CAM (computer-aided design and computer-aided manufacturing), BIM (building information modelling), GNSS (global navigation satellite system), GPS (United States global positioning system), GIS (geographic information systems, or GI sciences), sg (smart geometry).

10 Michael Batty, *Cities and Complexity: Understanding Cities with Cellular Automata, Agent-Based Models, and Fractals* (Cambridge, MA: The MIT Press, 2005). Michael Batty, *The New Science of Cities* (Cambridge, MA: The MIT Press, 2013).

11 Andrew Hudson-Smith, *Digitally Distributed Urban Environments: The Prospects for Online Planning* (PhD diss., London: University College London, Bartlett School of Graduate Studies, 1993), accessed January 3, 2016, http://www.casa.ucl.ac.uk/andy/thesis.pdf.

12 Carlo Ratti, Andres Sevtsuk, Burak Arikan, Assaf Biderman, Francesco Calabrese, Filippo Dal Fiore, Saba Ghole, Daniel Gutierrez, Sonya Huang, Sriram Krishnam, Justin Moe, Francisca Rojas, Najeeb Marc Tarazi (MIT SENSEable Cities Lab), "Real-Time Rome," Italian Pavilion, Venice Biennale of Architecture, September 10–November 19, 2006, accessed January 15, 2015, http://senseable.mit.edu/realtimerome/.

13 "Spaceship Earth: Observing Our Planet from Satellites," Customs House Sydney, May 23–July 20, 2014, accessed January 9, http://spaceship-earth-satellites.net/. "Spaceship Earth," September 3–6, 2015, Ars Electronica Center, Linz, accessed January 9, 2016, http://www.aec.at/postcity/en/spaceship-earth/.

14 European Commission, "Copernicus: Europe's Eyes on Earth," accessed January 11, 2016, http://www.copernicus.eu/. European Space Agency, "The Living Planet Program," accessed January 11, 2016, http://www.esa.int/Our_Activities/Observing_the_Earth/The_Living_Planet_Programme/ESA_s_Living_Planet_Programme. NASA, "Earth Observing System," accessed January 11, 2016, http://eospso.nasa.gov/. Digital Globe, "About Digital Globe," accessed January 11, 2016, https://www.digitalglobe.com/.

15 Resolution of satellite images varies with different types and qualities of remote sensing instruments. For example, ground sample distance (GSD, per image pixel) is approximately 30 metres resolution for older Landsat images, compared with 31 cm for Digital Globe's recent World View-3 imagery. Aerial photography offers much higher resolution (from one-off or fewer surveys than satellites).

16 Hanno-Walter Kruft, "Vitruvius and Architectural Theory in Antiquity," *A History of Architectural Theory from Vitruvius to the Present*, trans. Ronald Taylor, Elsie Callander and Antony Wood (London: Zwemmer and New York: Princeton Architectural Press, 1994), 24. Morris Hicky Morgan (trans.), *Vitruvius: The Ten Books on Architecture* (Boston, MA: Harvard University Press, 1960 edn), accessed January 11, 2016, http://www.gutenberg.org/cache/epub/20239/pg20239-images.html. Morgan translated the Latin "firmitas" to mean "durability".

17 Sigfried Giedion, José Luis Sert and Fernand Léger [1943], "Nine Points on Monumentality," in *Architektur und Gemeinschaft*, ed. Sigfried Giedion (Hamburg: Rowohlt, 1956), accessed January 11, 2016, http://designtheory.fiu.edu/readings/giedion_nine_points.pdf. "1. Monuments . . . are intended to outlive the period which originated them . . . "

18 Diller Scofidio and Renfro, "Blur Building," accessed December 6, 2015, http://www.dsrny.com/projects/blur-building.

19 Carlo Ratti Associati, "Digital Water Pavilion: Carlo Ratti Associati," *YouTube*, March 12, 2012, accessed January 11, 2016, https://www.youtube.com/watch?v=C5pw354oHIM.

20 Organization for Economic Co-operation and Development (International Futures Programme), *The Space Economy at a Glance 2007* (Paris: OECD, 2007 [updated edns 2011, 2014]).

21 Other virtual globes include Microsoft Bing, NASA World Wind and Esri ArcGIS. See Michael F. Goodchild, Huadong Guo, Alessandro Annoni, Ling Bian, Kees de Bie, Frederick Campbell, Max Craglia, Manfred Ehlersg, John van Genderen, Davina Jackson, Anthony J. Lewis, Martino Pesaresi, Gábor Remetey-Fülöpp, Richard Simpson, Andrew Skidmore, Changlin Wang and Peter Woodgate, "Next-generation Digital Earth," *Proceedings of the National Academy of Sciences*, June 21, 2012, 11089, accessed February 3, 2016, http://www.pnas.org/content/109/28/11088.full.pdf.

22 Goodchild et al., "Next-generation Digital Earth," 11089–11094.

23 NASA, "Blue Marble – Image of the Earth from Apollo 17 (7 December, 1972)," accessed January 12, 2016, https://www.nasa.gov/content/blue-marble-image-of-the-earth-from-apollo-17.

24 Office of Charles and Ray Eames, "Powers of Ten (1977)," *YouTube*, August 26, 2010, accessed January 12, 2016, https://www.youtube.com/watch?v=0fKBhvDjuy0.

25 Gore, *Earth in the Balance*, 357–358.

26 "Eratosthenes," *Wikipedia*, accessed January 10, 2016, https://en.wikipedia.org/wiki/Eratosthenes, also Cornell University Extra-Galactic Group, "Eratosthenes (276–195 B.C.)," *Astronomy 2201: The History of the Universe*, accessed January 15, 2016, http://www.astro.cornell.edu/academics/courses/astro201/eratosthenes.htm.

27 Morgan (trans.), *Vitruvius*.

28 Goodchild et al., "Next-generation Digital Earth," 11089–11090.

29 Group on Earth Observations, accessed January 16, 2016, https://www.earthobservations.org/index.php.

30 Intergovernmental Panel on Climate Change, "Reports," accessed January 15, 2016, https://www.ipcc.ch/publications_and_data/publications_and_data_reports.shtml.

31 United Nations Framework Convention on Climate Change, "About UNFCCC," *UN Climate Change Newsroom*, accessed January 15, 2016, http://newsroom.unfccc.int/about/.

32 Global Climate Observing System, "GCOS Essential Climate Variables," accessed December 8, 2015, https://www.wmo.int/pages/prog/gcos/index.php?name=EssentialClimateVariables.

33 Stephen Briggs, "Essential Climate Variables and Megatrends," video 1F in *Monitoring Climate from Space* online course (London: FutureLearn and European Space Agency, 2015), accessed January 12, 2016, https://www.futurelearn.com/courses/climate-from-space.

34 NASA, "Catalog of Earth Satellite Orbits," *Earth Observatory*, accessed December 14, 2015, http://earthobservatory.nasa.gov/Features/OrbitsCatalog/. NASA, "What Are Passive and Active Sensors?," October 15, 2012, accessed December 14, 2015, https://www.nasa.gov/directorates/heo/scan/communications/outreach/funfacts/txt_passive_active.html.

35 Fred Pearce, "Greenwash: The Dream of the First Eco-City Was Built on a Fiction," *The Guardian*, April 23, 2009, accessed January 12, 2016, http://www.theguardian.com/environment/2009/apr/23/greenwash-dongtan-ecocity.

36 Amy Pollock, "French Architect Designs a Sustainable Vertical City to be Installed in the Sahara Desert," *Reuters* (US), July 21, 2015, accessed January 8, 2016, http://www.reuters.com/article/us-sahara-city-idUSKCN0PG16G20150721.

37 Rachel Armstrong, *Vibrant Architecture: Matter as a Co-Designer of Living Structures* (Berlin, Warsaw: De Gruyter Open, 2015), accessed January 12, 2016, http://www.degruyter.com/view/product/448453.

38 Danny Hudson, "Magnus Larsson Sculpts the Saharan Desert with Bacteria," *Designboom*, June 24, 2013, accessed January 8, 2016, http://www.designboom.com/architecture/magnus-larsson-sculpts-the-saharan-desert-with-bacteria/.

39 European Union, *Official Journal of the European Union*, 50:108 (2007), EUR-Lex, accessed December 7, 2015, http://eur-lex.europa.eu/legal-content/EN/ALL/?uri=OJ:L:2007:108:TOC.

40 European Commission, "Data Specifications," *Infrastructure for Spatial Information in the European Community (INSPIRE)*, accessed December 8, 2015, http://inspire.ec.europa.eu/index.cfm/pageid/2/list/7.

41 Jackson and Simpson (eds.), "Essential Climate Variables," *D_City*, 8, and "INSPIRE: A benchmark program," *D_City*, 91.

42 Jackson and Simpson (eds.), "Virtual Nations and Networks," *D_City*, 82–95.

43 Martin Laplante, "Evidence-Based Urban Planning," *Planetizen*, November 1, 2010, accessed January 12, 2016, http://www.planetizen.com/node/46699.

44 International Standards Organization, "ISO 37120:2014," accessed January 15, 2016, http://www.iso.org/iso/catalogue_detail?csnumber=62436. World Council on City Data, "Created by Cities, for Cities," accessed January 15, 2016, http://www.dataforcities.org/wccd/. Global City Indicators Facility, accessed January 15, 2016, http://www.cityindicators.org/Default.aspx.

45 The Global Urban Observatory network was launched at UN-Habitat's *Habitat II* conference on human settlements in Istanbul, 3–14 June, 1996. UN-Habitat, "Global Urban Observatory (GUO)," accessed January 16, 2016, http://unhabitat.org/urban-knowledge/global-urban-observatory-guo/.

46 Richard Saul Wurman, *The City: Form and Intent: Being a Collection of the Plans of Fifty Significant Towns and Cities All to the Scale 1:1440* (Raleigh, NC: University of North Carolina at Raleigh, 1963).

47 Richard Saul Wurman, *Design Quarterly 80: Making the City Observable* (Minneapolis, MN: Walker Art Center and Cambridge, MA: MIT Press, 1971), 75–76.

48 Wurman, *Making the City Observable*, "Author's Introduction," 6.

49 Societal benefit areas in GEO's new strategic plan are more goal oriented than the SBAs originally defined in 2005. They now are biodiversity and ecosystem sustainability, disaster resilience, energy and mineral resources management, food security and sustainable agriculture, infrastructure and transportation management, public health surveillance, sustainable urban development and water resources management. *GEO Strategic Plan 2016–2025: Implementing GEOSS*. (Geneva: Group on Earth Observations, 2015), 8–10.

50 Winy Maas, Arie Graafland, Arjen van Susteren, Arthur van Bilsen, Brent Batstra, Camillo Pinilla, Daniel Dekkers and collaborators, *SpaceFighter: The Evolutionary City (Game:)* (Rotterdam: MVRDV, Delft School of Design, Berlage Institute, MIT, 2006).

51 Gehry Partners recently sold its building and cities modelling subsidiary, Gehry Technologies, to Trimble. Anna Winston, "Frank Gehry's Technology Company Bought by Sketch Up Owners," *Dezeen*, September 8, 2014, accessed January 12, 2016, http://www.dezeen.com/2014/09/08/trimble-buys-frank-gehry-technologies/.

52 Terri Peters and Brady Peters, *Inside Smartgeometry: Expanding the Architectural Possibilities of Computational Design* (London: Wiley, 2013).

53 The five groups (of computer-aided architects and designers) aligned with the Netherlands-based CAAD Futures Foundation are eCAADe (Europe), ACADIA (North America), SIGRADI (Latin America), CAADRIA (Asia-Oceania) and ASCAAD (Arab nations). They share the CuminCAD online database of articles.

54 University College London's Centre for Advanced Spatial Analysis developed the "Virtual London" 3D online model as a platform for what-if scenarios, including flooding of the Thames, and has new funding to update it. Hudson-Smith, *Digitally Distributed Urban*

Environments, 2003. Batty, *Cities and Complexity*, 2005; Batty, *New Science of Cities*, 2013.

55 UCL's Space Syntax unit was founded on principles of computationally analysing urban movement patterns. Bill Hillier, *Space Is the Machine: A Configural Theory of Architecture* (London: Cambridge University Press, 1996).

56 Rajendra K. Pachauri, Leo Meyer et al, *Climate Change 2014: Fifth Assessment Report of the Intergovernmental Planel on Climate Change* (Geneva: IPCC/World Meteorological Organization, 2014), 6–17, accessed January 12, 2016, https://www.ipcc.ch/pdf/assessment-report/ar5/syr/SYR_AR5_FINAL_full.pdf.

57 Bob Bishop, *Towards a Globally Focused Earth Simulations Centre* (Geneva: ICES Foundation, 2013), 7.

58 ICES Foundation. "International Centre for Earth Simulation," accessed December 15, 2015, http://www.icesfoundation.org.

59 Bishop, *Earth Simulations Centre*, 7.

60 Geoscience Australia, "Australian Geospatial Data Cube," accessed December 9, 2015, http://www.datacube.org.au/.

61 The Data Cube is supported by the Committee on Earth Observation Sciences (CEOS) and the Group on Earth Observations (GEO), both supported by most European and global spatial science and United Nations organisations.

62 Martino Pesaresi, Xavier Blaes, Daniele Ehrlich, Stefano Ferri, Lionel Guegen, Fernand Haag, Martina Halkia, Johannes Heinzel, Mayeul Kauffmann, Thomas Kemper, Georgios K. Ouzounis, Marco Scavazzon, Pierre Soille, Vasileios Syrris and Luigi Zanchetta, *A Global Human Settlement Layer from Optical High Resolution Imagery: Concept and First Results* (Ispra: European Commission JRC Scientific and Policy Report, 2012), 5–7, 12–14, 25–29, 97–99.

63 M. Pesaresi, D. Ehrlich, S. Ferri, A. Florczyk, S. Freire, F. Haag, M. Halkia, A.M. Julea, T. Kemper and P. Soille, "Global Human Settlement Analysis for Disaster Risk Reduction," *International Archives of Photogrammetry and Remote Sensing, Spatial Information Sciences* XL-7/W3 (2015): 837–843. M. Pesaresi, H. Guo, X. Blaes, D. Ehrlich, S. Ferri, L. Gueguen, M. Halkia, M. Kauffmann, T. Kemper, L. Lu, M.A. Marin-Herrera, G.K. Ouzounis, M. Scavazzon, P. Soille, V. Syrris and L. Zanchetta, "A Global Human Settlement Layer from Optical HR/VHR RS Data: Concept and First Results," *IEEE Journal of Selected Topics in Applied Earth Observations and Remote Sensing* 6:5 (October 2013): 2102–2131.

64 Carl Steinitz, *A Framework for Geodesign: Changing Geography by Design* (Redlands, CA: Esri Press, 2012).

65 Steinitz, *Framework for Geodesign*, 59. Andrew Crooks, "Book Review: A Framework for Geodesign", *GIS and Agent-Based Modelling*, December 11, 2013, accessed February 3, 2016, http://www.gisagents.org/2013/12/book-review-framework-for-geodesign.html.

66 First publication of the term "licht architektur" has been credited to Joachim Teichmüller in a 1927 edition of the German trade journal, *Licht und Lampe*. "Architecture of the Night," *Wikipedia*, accessed January 15, 2016, https://www.google.com.au/search?q=teichmuller+licht+und+lampe&ie=utf-8&oe=utf-8&gws_rd=cr&ei=f0uYVvCNKaK7mwWGk6mgCg. Werner Oechslin, "Lichtarchitektur," *Moderne Architektur in Deutschland 1900 Bis 1950: Expressionismus und Neue Sachlichkeit*, ed. Vittorio Magnagno Lampugnani and Romana Schneider (Stuttgart: Hatje, 1994), 117.

67 Davina Jackson coined the term "smart light cities" to help strengthen the "smart cities" movement promulgated by MIT professor William J. Mitchell. This term was used in successful proposals to the New South Wales (Australia) and Singapore governments for three "Smart Light" festivals developed by Mary-Anne Kyriakou and collaborators, 2009–2012. "Smart Light Sydney", accessed January 15, 2016, http://smartlightsydney.org. Davina Jackson, "Cities of Cool Light," *SuperLux: Smart Light Art, Design and Architecture for Cities* (London: Thames & Hudson, 2015), 8–9. *SuperLux*, accessed January 15, 2016, http://superlux.org.

68 Marcos Novak has promoted the term "trans-architecture" since the mid-1990s and leads the TransLAB at the University of California Santa Barbara. "Novak, Marcos: Trans-Architecture, 1994," *Fen-Om Theory*, accessed January 15, 2016, http://www.fen-om.com/theory/theory12.pdf. "Novak, Marcos: The Meaning of Trans-Architecture, 1996," *Fen-Om Theory*, accessed January 15, 2016, http://issuu.com/salberti/docs/meaning-of-transarchitecture.

69 Media Architecture Institute, accessed January 15, 2016, http://www.mediaarchitecture.org/.
70 International Society for Presence Research, accessed January 15, 2016, https://ispr.info/.
71 Scientists using QED are inspired by Richard Feynman's lectures in Auckland (1979) and Los Angeles (1983). Richard Feynman, *QED: The Strange Theory of Light and Matter*, ed. Ralph Leighton (New York: Princeton University Press, 1985). Ladislav Szantó, "'Principles' and 'Holography,'" *The Strange Theory of Light: Animation of Feynman Pictures Light by QED*, accessed January 15, 2016, http://qed.wikina.org/principles/.
72 Joachim Krausse and Claude Lichtenstein (eds.), "Chronology," *Your Private Sky: R. Buckminster Fuller: The Art of Design Science* (Baden: Lars Müller, 1999), 26–39. Richard Buckminster Fuller and Kiyoshi Kuromiya, *Cosmography: A Posthumous Scenario for the Future or Humanity* (New York: Macmillan Publishing Co., 1992).
73 Richard Buckminster Fuller, "World Planning," *The Buckminster Fuller Reader*, ed. James Meller (London: Jonathon Cape, 1970), 363–364.
74 Feynman, *QED*.
75 The first formal organisation of space architects is the Space Architecture Technical Committee (SATC) of the American Institute of Aeronautics and Astronautics (AAA), accessed February 16, 2016, http://spacearchitect.org/about-us/. "Space Architecture", *Wikipedia*, accessed February 16, 2016, https://en.wikipedia.org/wiki/Space_architecture.
76 Hudson-Smith, *Digitally-Distributed Urban Environments*, 109.
77 Hudson-Smith, *Digitally-Distributed Urban Environments*, 109.

Bibliography

"Architecture of the night." *Wikipedia*. Accessed January 15, 2016. https://www.google.com.au/search?q=teichmuller+licht+und+lampe&ie=utf-8&oe=utf-8&gws_rd=cr&ei=f0uYVvCNKaK7mwWGk6mgCg.

Armstrong, Rachel. *Vibrant Architecture: Matter as a Co-Designer of Living Structures*. Berlin, Warsaw: De Gruyter Open, 2015. Accessed January 12, 2016. http://www.degruyter.com/view/product/448453.

Batty, Michael J. *Cities and Complexity: Understanding Cities with Cellular Automata, Agent-Based Models, and Fractals*. Cambridge, MA: The MIT Press, 2005.

Batty, Michael J. *The New Science of Cities*. Cambridge, MA: The MIT Press, 2013.

Berggren, J. Lennart and Alexander Jones. *Ptolemy's Geography: An Annotated Translation of the Theoretical Chapters*, 57–59. Princeton, NJ and Oxford: Princeton University Press, 2000.

Bishop, Robert. *Towards a Globally Focused Earth Simulations Centre*, 7. Geneva: ICES Foundation, 2013.

Briggs, Stephen. Quoted in "Essential Climate Variables and Megatrends." Video 1F, *Monitoring Climate from Space* (online course). Produced by FutureLearn and European Space Agency, 2015. Accessed January 12, 2016. https://www.futurelearn.com/courses/climate-from-space.

Carlo Ratti Associati. "Digital Water Pavilion: Carlo Ratti Associati". *YouTube*, March 12, 2012. Accessed January 11, 2016. https://www.youtube.com/watch?v=C5pw354oHIM.

Charles and Ray Eames (Office). "Powers of Ten (1977)." *YouTube*, August 26, 2010. Accessed January 12, 2016. https://www.youtube.com/watch?v=0fKBhvDjuy0.

Cornell University Extra-Galactic Group. "Eratosthenes (276–195 B.C.)." *Astronomy 2201: The History of the Universe*. Accessed January 15, 2016. http://www.astro.cornell.edu/academics/courses/astro201/eratosthenes.htm.

Crooks, Andrew. "Book Review: A Framework for Geodesign." *GIS and Agent-Based Modelling*. December 11, 2013. Accessed February 3, 2016. http://www.gisagents.org/2013/12/book-review-framework-for-geodesign.html.

Digital Globe. "About Digital Globe." Accessed January 11, 2016. https://www.digitalglobe.com/.

Diller Scofidio and Renfro. "Blur Building." Accessed December 6, 2015. http://www.dsrny. com/projects/blur-building.

"Eratosthenes." *Wikipedia*. Accessed January 10, 2016. https://en.wikipedia.org/wiki/ Eratosthenes.

European Commission. "Copernicus: Europe's Eyes on Earth." Accessed January 11, 2016. http://www.copernicus.eu/.

European Commission. "Data Specifications." *Infrastructure for Spatial Information in the European Community (INSPIRE)*. Accessed December 8, 2015. http://inspire.ec.europa.eu/ index.cfm/pageid/2/list/7.

European Space Agency. "The Living Planet Programme." Accessed January 11, 2016. http:// www.esa.int/Our_Activities/Observing_the_Earth/The_Living_Planet_Programme/ ESA_s_Living_Planet_Programme.

European Union. *Official Journal of the European Union*. 50: 108 (April 25, 2007). Accessed December 7, 2015. http://eur-lex.europa.eu/legal-content/EN/ALL/?uri=OJ:L:2007:108:TOC.

Feynman, Richard. *QED: The Strange Theory of Light and Matter*. Edited by Ralph Leighton. New York: Princeton University Press, 1985.

Fuller, Richard Buckminster. *An Operating Manual for Spaceship Earth*. New York: E.P. Dutton, 1968.

Fuller, Richard Buckminster. "World Planning." *The Buckminster Fuller Reader*. Edited by James Meller. London: Jonathon Cape, 362–368, 1970.

Fuller, Richard Buckminster and Kiyoshi Kuromiya. *Cosmography: A Posthumous Scenario for the Future of Humanity*. New York: Macmillan Publishing Co., 1992.

Geoscience Australia. "Australian Geospatial Data Cube." Accessed December 9, 2015. http:// www.datacube.org.au/.

Giedion, Sigfried, José Luis Sert and Fernand Léger. "Nine Points on Monumentality." *Architektur und Gemeinschaft*. Edited by Sigfried Giedion. Hamburg: Rowohlt, 1956. Accessed January 11, 2016. http://designtheory.fiu.edu/readings/giedion_nine_points.pdf.

Global City Indicators Facility. Accessed January 15, 2016, http://www.cityindicators.org/ Default.aspx.

Global Climate Observing System. "GCOS Essential Climate Variables." Accessed December 8, 2015. https://www.wmo.int/pages/prog/gcos/index.php?name=EssentialClimateVariables.

Goodchild, Michael F., Huadong Guo, Alessandro Annoni, Ling Bian, Kees de Bie, Frederick Campbell, Max Craglia, Manfred Ehlersg, John van Genderen, Davina Jackson, Anthony J. Lewis, Martino Pesaresi, Gábor Remetey-Fülöpp, Richard Simpson, Andrew Skidmore, Changlin Wang and Peter Woodgate. "Next-Generation Digital Earth." *Proceedings of the National Academy of Sciences* (June 21, 2012): 11089–11094. Accessed February 3, 2016. http://www.pnas.org/content/109/28/11088.full.

Gore, Al. *Earth in the Balance: Ecology and the Human Spirit*. New York: Houghton Mifflin Harcourt, 1992.

Group on Earth Observations. *GEO Strategic Plan 2016–2025: Implementing GEOSS*. Geneva: GEO, 2015.

Hillier, Bill. *Space Is the Machine: A Configural Theory of Architecture*. London: Cambridge University Press, 1996.

Hoehler, Sabine. *Spaceship Earth in the Environmental Age, 1960–1990*. London: Routledge, 2015.

Hudson, Danny. "Magnus Larsson Sculpts the Saharan Desert with Bacteria." *Designboom*, June 24, 2013. Accessed January 8, 2016. http://www.designboom.com/architecture/ magnus-larsson-sculpts-the-saharan-desert-with-bacteria/.

Hudson-Smith, Andrew. *Digitally Distributed Urban Environments: The Prospects for Online Planning* (PhD diss.). London: University College London, Bartlett School of Graduate Studies, 1993.

ICES Foundation. "International Centre for Earth Simulation." Accessed December 15, 2015. http://www.icesfoundation.org.

Intergovernmental Panel on Climate Change. "Reports." Accessed January 15, 2016, https://www.ipcc.ch/publications_and_data/publications_and_data_reports.shtml.

International Society for Presence Research. Accessed January 15, 2016. https://ispr.info/.

International Standards Organization. "ISO 37120:2014." Accessed January 15, 2016. http://www.iso.org/iso/catalogue_detail?csnumber=62436.

Jackson, Davina. "Cities of Cool Light." *SuperLux: Smart Light Art, Design and Architecture for Cities*, 8–9. London: Thames & Hudson, 2015.

Jackson, Davina and Richard Simpson, eds. *D_City: Digital Earth | Virtual Nations | Data Cities [Global Futures for Environmental Planning]*, 8, 82–95. Ebook 2012. http://dcitynetwork.net/manifesto. Accessed January 11, 2016. Print edn Sydney: DCity and Geneva: GEO, 2013.

Krausse, Joachim and Claude Lichtenstein, eds. "Chronology." *Your Private Sky: R. Buckminster Fuller: The Art of Design Science*, 26–39. Baden, CH: Lars Müller, 1999.

Kruft, Hanno-Walter. (1985). "Vitruvius and Architectural Theory in Antiquity." *A History of Architectural Theory from Vitruvius to the Present*, 21–29. Translated by Ronald Taylor, Elsie Callander and Antony Wood. London: Zwemmer and New York: Princeton Architectural Press, 1994.

Laplante, Martin. "Evidence-Based Urban Planning." *Planitizen*, November 1, 2010. Accessed January 12, 2016. http://www.planetizen.com/node/46699.

Maas, Winy, Arie Graafland, Arjen van Susteren, Arthur van Bilsen, Brent Batstra, Camillo Pinilla, Daniel Dekkers and collaborators. *SpaceFighter: The Evolutionary City (Game)*. Rotterdam: MVRDV, Delft School of Design, Berlage Institute, MIT, 2006.

Media Architecture Institute. Accessed January 15, 2016. http://www.mediaarchitecture.org/.

Mitchell, William J. *City of Bits: Space, Place and the Infobahn*, 24. Cambridge, MA: The MIT Press, 1995.

Morgan, Morris Hicky, trans. *Vitruvius: The Ten Books on Architecture*. Boston, MA: Harvard University Press, 1960 edn. Accessed January 11, 2016. http://www.gutenberg.org/cache/epub/20239/pg20239-images.html.

NASA. "Blue Marble – Image of Earth from Apollo 17 (7 December, 1972)." Accessed January 12, 2016. https://www.nasa.gov/content/blue-marble-image-of-the-earth-from-apollo-17.

NASA. "Catalog of Earth Satellite Orbits." *Earth Observatory*. Accessed December 14, 2015. http://earthobservatory.nasa.gov/Features/OrbitsCatalog/.

NASA. "Earth Observing System." Accessed January 11, 2016. http://eospso.nasa.gov/.

NASA. "What Are Passive and Active Sensors?" October 15, 2012. Accessed December 14, 2015. https://www.nasa.gov/directorates/heo/scan/communications/outreach/funfacts/txt_passive_active.html.

Novak, Marcos. "Novak, Marcos: The Meaning of Trans-Architecture, 1996." Reprinted in *Fen-Om Theory*. Accessed January 15, 2016. http://issuu.com/salberti/docs/meaning-of-transarchitecture.

Novak, Marcos. "Novak, Marcos: Trans-Architecture, 1994." Reprinted in *Fen-Om Theory*. Accessed January 15, 2016. http://www.fen-om.com/theory/theory12.pdf.

Oechslin, Werner. "Lichtarchitektur." *Moderne Architektur in Deutschland 1900 Bis 1950: Expressionismus and Neue Sachlichkeit*, 117. Edited by Vittorio Magnagno Lampugnani and Romana Schneider. Stuttgart: Hatje, 1994.

Organization for Economic Cooperation and Development. *The Space Economy at A Glance*. Paris: OECD, 2007, 2011, 2014.

Pachauri, Rajendra K., Leo Meyer et al., *Climate Change 2014: Fifth Assessment Report of the Intergovernmental Planel on Climate Change*, 6–17. Geneva: IPCC/World Meteorological Organization, 2014. Accessed January 12, 2016. https://www.ipcc.ch/pdf/assessment-report/ar5/syr/SYR_AR5_FINAL_full.pdf.

Pearce, Fred. "Greenwash: The Dream of the First Eco-City Was Built on a Fiction." *The Guardian*, April 23, 2009. Accessed January 12, 2016. http://www.theguardian.com/environment/2009/apr/23/greenwash-dongtan-ecocity.

Pesaresi, Martino, Xavier Blaes, Daniele Ehrlich, Stefano Ferri, Lionel Guegen, Fernand Haag, Martina Halkia, Johannes Heinzel, Mayeul Kauffmann, Thomas Kemper, Georgios K. Ouzounis, Marco Scavazzon, Pierre Soille, Vasileios Syrris and Luigi Zanchetta. *A Global Human Settlement Layer from Optical High Resolution Imagery: Concept and First Results*, 5–7, 12–14, 25–29, 97–99. Ispra: European Commission JRC Scientific and Policy Report, 2012.

Pesaresi, M., D. Ehrlich, S. Ferri, A. Florczyk, S. Freire, F. Haag, M. Halkia, A.M. Julea, T. Kemper and P. Soille. "Global Human Settlement Analysis for Disaster Risk Reduction." *International Archives of Photogrammetry and Remote Sensing, Spatial Information Sciences* XL-7/W3 (2015): 837–843.

Pesaresi, M., H. Guo, X. Blaes, D. Ehrlich, S. Ferri, L. Gueguen, M. Halkia, M. Kauffmann, T. Kemper, L. Lu, M.A. Marin-Herrera, G.K. Ouzounis, M. Scavazzon, P. Soille, V. Syrris and L. Zanchetta. "A Global Human Settlement Layer from Optical HR/VHR RS Data: Concept and First Results." *IEEE Journal of Selected Topics in Applied Earth Observations and Remote Sensing* 6:5 (2013): 2102–2131.

Peters, Terri and Brady Peters. *Inside Smartgeometry: Expanding the Architectural Possibilities of Computational Design.* London: Wiley, 2013.

Pollock, Amy. "French Architect Designs a Sustainable Vertical City to be Installed in the Sahara Desert." *Reuters* (US), July 21, 2015. Accessed January 8, 2016. http://www.reuters.com/article/us-sahara-city-idUSKCN0PG16G20150721.

Poltorzycki. "Dongtan Eco City: History 60c Project." *YouTube*, April 25, 2014. Accessed December 9, 2015. https://www.youtube.com/watch?v=sJ6WW0f_xR8.

Ratti, Carlo, Andres Sevtsuk, Burak Arikan, Assaf Biderman, Francesco Calabrese, Filippo Dal Fiore, Saba Ghole, Daniel Gutierrez, Sonya Huang, Sriram Krishnam, Justin Moe, Francisca Rojas, Najeeb Marc Tarazi (MIT SENSEable Cities Lab). "Real-Time Rome," Italian Pavilion, Venice Biennale of Architecture, September 10–November 19, 2006, Accessed January 15, 2015, http://senseable.mit.edu/realtimerome/.

Smart Light Sydney. Accessed January 15, 2016. http://smartlightsydney.org.

"Space Architecture." *Wikipedia*. Accessed February 16, 2016. https://en.wikipedia.org/wiki/Space_architecture.

Space Architecture Technical Committee (SATC) of the American Institute of Aeronautics and Astronautics. "Spacearchitect.org." Accessed February 16, 2016. http://spacearchitect.org/about-us/.

"Spaceship Earth." *Ars Electronica Center*. Accessed January 9, 2016. http://www.aec.at/postcity/en/spaceship-earth/.

Spaceship Earth: Observing Our Planet from Satellites. Accessed January 11, 2016. http://spaceship-earth-satellites.net/.

"Spaceship Earth." *Wikipedia*. Accessed January 9, 2016. https://en.wikipedia.org/wiki/Spaceship_Earth.

Steinitz, Carl. *A Framework for Geodesign: Changing Geography by Design*. Redlands, CA: Esri Press, 2012.

SuperLux. Accessed January 15, 2016. http://superlux.org.

Szantó, Ladislav. "'Principles' and 'Holography'." *The Strange Theory of Light: Animation of Feynman Pictures Light by QED*. Accessed January 15, 2016. http://qed.wikina.org/principles/.

UN-Habitat. "Global Urban Observatory (GUO)." Accessed January 16, 2016. http://unhabitat.org/urban-knowledge/global-urban-observatory-guo/.

United Nations Conference on Trade and Development. *Economic Development in Africa Report 2015: Unlocking the Potential of Africa's Services Trade for Growth and Development*, 38. New York and Geneva: United Nations, 2015.

United Nations Framework Convention on Climate Change. "About UNFCCC." *UN Climate Change Newsroom*. Accessed January 15, 2016. http://newsroom.unfccc.int/about/.

Winston, Anna. "Frank Gehry's Technology Company bought by SketchUp Owners." *Dezeen*, September 8, 2014. Accessed January 12, 2016. http://www.dezeen.com/2014/09/08/trimble-buys-frank-gehry-technologies/.

Wolfram, Stephen. *A New Kind of Science*, 7–16. Champaign, IL: Wolfram Media, 2002.

World Council on City Data. "Created by Cities, for Cities." Accessed January 15, 2016, http://www.dataforcities.org/wccd/.

Wurman, Richard Saul. *Design Quarterly 80: Making the City Observable*. Edited by Mildred Friedman. Minneapolis, MN: Walker Art Center and Cambridge, MA: MIT Press, 1971.

Wurman, Richard Saul. *The City: Form and Intent: Being a Collection of the Plans of Fifty Significant Towns and Cities All to the Scale 1:1440*. Raleigh, NC: University of North Carolina at Raleigh, 1963.

12 Leveraging nature to envision (functional) space

An architecture of machinic abduction

Tim Ireland

Introduction

The word "sight" is typically associated with the visual sense of seeing, or observing, with one's eyes. At base, the word is fundamentally concerned with "perception" in terms of perceiving what is, is to come or might be, which has an altogether conceptually different connotation. As a noun, "sight" pertains to the capacity to realise, grasp and comprehend, or to perceive in a manner which is not specifically oriented to seeing with one's eyes. Shaman and holy persons are one example of which people turn to for *in*sight and discernment of events they cannot "see" for themselves.[1] We tend to refer to our sensorial capacity as composed of five senses: vision, hearing, smell, touch and taste. This is, however, a compounded view of our sensorial capacities, which is prompted by the anthropocentric view of existence.[2] The manner in which this overriding view has shaped our understanding of existence has fuelled the pre-eminence of our visual sense over all others and informed technological and cultural (specifically Western society) progress in a manner which is focused on the ability to see visually.[3]

The Renaissance representation of space was fundamentally grounded in an ideological conception of the world. The vanishing point and the meeting of parallel lines "at infinity" evokes a God-like standpoint, determining a representation at once both intellectual and visual, which promoted the primacy of the gaze in a kind of "logic of visualisation".[4] Coupled with modern physics (if we negate certain relativistic theories) space was qualified as continuous, isotropic, homogenous, finite or infinite and fundamentally a matter, given credence by visually oriented techniques of representation, controlled by means of its visual articulation. The obsession with perspective is entrenched in the perceived superiority of sight and the significance of the image, which was followed, and accentuated, by advances in photography and film. What the image offers is the articulation of space (as visual) itself.

> The claim is that space can be shown by means of space itself. Such a technique (also known as tautology) uses and abuses a familiar technique that is as easy to abuse as it is to use – namely a shift from the part to the whole: metonymy. Take images, for example: photographs, advertisements, films. Can images of this kind really be expected to expose errors concerning space? Hardly, where there is error or illusion, the image is more likely to secrete it and reinforce it than to reveal it.[5]

The emphasise on the visual manifestation of space has in turn determined how architects comprehend the correlation between society and the environment and, ultimately, influenced if not dictated the manner in which they perceive, manufacture and develop their ideas into built form. The objectified notion of space – as visually prescribed – prevails in contemporary techniques of representation[6] with, for example, computation being utilised to fuel hyper-real visualisation of projects[7] or produce mute virtual reality environments that visually mimic the physical world.[8]

In this chapter, a significantly different approach is proposed involving first, a challenge to the prominence of sight in our conceptualisation of space generally, followed by the argument that the manner in which computation is employed as a tool in design needs to be reconsidered to account for this challenge to the limitations of the visual. It suggests that the most recent potentialities offered by computation in the field of spatial design conceptually draw on issues of semiosis and the abductive (non-anthropocentric and non-materialistic) nature of living systems that free our understanding of space and the role of computation in its creation and representation. In doing so, it will be suggested that the next steps for the role of technologies in the evolution of architecture may not be visual at all, but rather more fundamental, involving an ever-greater potential to mimic, learn and engage with natural systems at a level much deeper than any human sense, most of all the purely visual.

Technologies of visualisation – moving beyond sight

The argument that architecture is dominated by a concern for the visual is long held,[9] along with the view that architects should account for a wider sensorial domain in the artefacts they create.[10] Technology enables us to manipulate and engage with sensorial abilities beyond the five senses,[11] and architects now have the potential to capitalise on these abilities to enrich their design thinking and artefact making. Doing so requires a shift in how we conceive being-in-the-world and understand the correspondence between ourselves, as organisms, and the environment. Questioning entrenched views of mind, representation, space, intelligence, time and information, among others, and what it is to be human are allowing us to overcome the Cartesian split and establish a holistic, unified approach to designing architectural scenarios that articulate the coupling between an organism and its environment, which enriches rather than fractures the correlation between people, society, built environment and natural world. To do this, it is necessary to employ a naturalised conception of space,[12] drawing on biological theory to provoke a decentralised, distributed and emergent approach to architecture and consider how a reconceptualisation of how we use technology is opening up the potential of the body's sensorial capacities, which extends the typically held view that we cognise by means of five senses alone and in particular that vision is the sense above others. Such an approach calls for a radical shift in our use and application of technology. Artists and designers are today utilising technology in a more systemic manner, prompting and leveraging the tendencies of systemic assemblies in directions that cajole and provoke the creative propensity of these systems, thus perturbing their habitual existence into new directions. In other words, the pattern-recognition/making predilections of these systems are disturbed and thereby influenced and driven into new regimes of behaviour and pattern-making tendencies. Consequently, the dominance of the visual is challenged as these designers cogitate sensorial engagement, and in exploring a more distributed and emergent manner, they edge beyond the visually oriented dictum. Such efforts are, however, hampered by our inherent reliance on the visual. The sense of sight played a significant role in our evolution as a species, and

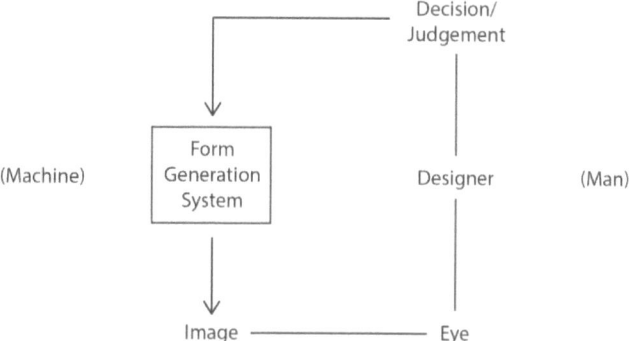

Figure 12.1 Interaction between man and machine, the 'see and decide relationship', by Broughton, Tan and Coates (1997); redrawn by the author[13]

we are thus a very visual mammal. Consequently, technology is dominated by visual interface. Computers, for example, having become ubiquitous in design, are reliant on their graphic output, because whatever one does, typically, with a computer is conditioned by reading what one sees on the screen (see Fig. 12.1). The avant-garde architect today is not a parametricist[14] but a philosopher, system designer, computer programmer, roboticist and material and biological scientist. A better idiom perhaps would be an entrepreneur of technological capacities who invents devices, or scenarios, that extend conscious cognitive abilities to open up potentialities to thereby tinker and interfere with preconditioned conceptions of what architecture is and what it is to be an architect. In other words, to produce an architecture of *in*sight that provokes, questions and explores perceived limitations of creativity.

Visualising space: The inside-outside problematic

Ever since classical science mathematised space, there has been an inclination to redefine what we understand space to be: to break its objectification. "It is because we think of space as Euclidean that its elusiveness to perception is so striking".[15] Space is a property of the world, which cannot be rationalised in the same way an object can. As Bruno Latour points out "physical space is not an object".[16] "Physical space" is an oxymoron pertaining to particular quantitative attributes of objects and (outside of everyday language) should be disregarded, or at least recognised for what it is: the absolute expression of objective properties through geometric and mathematical rules. The representation of space is conditional on the technique used, delineating a cognitive loop between representation and comprehension that narrows the conception of space to a direct reference to that which is perceived. In other words the method of representation, which is usually visual, determines the perception of that represented, or vice versa. Mathematical space is an abstraction of experience, being a set of rules by which experience may be replicated.[17] Spatial problems are inherently situated in the world, which we manage and solve within the confines of geometry. This is the strength of geometry: that it states general laws about geometrical objects and scenarios that we can then apply back through visual manifestation to the real world.

George Berkeley (1685–1753) put forward a notion of space, in direct opposition to absolute space, based on sensorial interaction with the world.[18] He emphasised the significance of communication – that we are a part of the world with which we interact exteroceptively and introeceptively. Interaction is thus mediated, consequent of the correlation between external and internal stimuli. However, Berkeley's proposition of a sensorial conception of space was based on sight and touch being the two senses par excellence, with sight being held in the highest regard. He also emphasised the significance of motion, stating,

> Supposing all the world to be annihilated besides my own body, I say there still remains pure space, thereby nothing else is meant but only that I conceive it possible for the limbs of my body to be moved on all sides without the least resistance; but if that, too, were annihilated, then there could be no motion, and consequently no space.[19]

"Life is but motion" claimed Thomas Hobbes (1588–1679). I may remain physically still but blood continues to traverse my veins, my heart beats, my lungs inhale and exhale and the cells (the basic structural, functional, and biological unit of all known living organisms) constituting my being replicate and die. Life is motion because any matter above absolute zero will be in motion, so since energy ceases at this ground state so does motion.

At the beginning of the twentieth century, as Peter Weibel explains, the perception of motion and the mechanisms of the perception of form overtook artists' preoccupation with exploring colour and its composition and effect on the eye.[20] Cubism, Futurism (among others) and the experimental photography and film of the 1920s are evidence of this turn. This shift of attention, he explains, towards the phenomenon of movement was compelled by the emergence of machines and the advancement of gyratory technology. Moholy-Nagy's *Light-Space Modulator* is an exemplar of this shift in architectural focus to emphasise architecture as creation in the mastery of the space experience. Influenced by developments in psychology, architects began to take a scientific approach to production and in theorising in order to validate their work.

Distinguishing a difference

The Estonian theoretical biologist Jacob von Uexküll emphasised that we see the world in a way that is useful to us, not as it is, by establishing patterns and finding relationships between things with which we associate some significance. He describes a functional niche (or way of life) which each organism defines – fitting the world to itself by integrating qualities of the world into its own coherent system. Reiterating Berkeley's claim (regarding the significance of movement, for his sensorial conception of space, as consequent to the mediation of internal and external stimuli) Uexküll claims, "If we recognise in what-lies-outside-ourselves the possibility of movement, then space as the connection of this possibility with the planes of direction, will be true "form", – namely, possibility and law."[21] It is through the capacity to sense that an organism is affected by, and thereafter affects, the world as a result of variations in the environment which it perceives and acts on by means of its ability to identify and distinguish differences.[22] The notion of difference, which was central to Gregory Bateson's ecological standpoint, is resonated in William Mitchell's definition of architecture as an "art of distinctions", resolving boundaries between categories around which differences are recognised. "When such distinctions are made an amorphous world is transformed into a world that has distinct parts organised in some particular way".[23] The manner in which we conceive the spatial correlation between one thing and another is fundamental to how we think about and thereby represent space.[24] This cognising determines how we approach the organisation of our spatial environment – and this leads to a distinction in how we construct artefacts.

Artefact making: Computation and coding

Life and architecture are both concerned with artefact making, but there is a significant disparity between the two. Frederick Kiesler recognised the difference between how nature and human-kind builds, stating, "Nature builds by cell division towards continuity whilst man can only build by joining together into a unique structure without continuity".[25] The emphasis behind Kiesler's distinction is that architects tend to make things through brute force (connecting parts together to form a whole), whereas nature tends to produce through a process of continuous construction whereby parts merge, overlap and conjoin. The latter is the product of self-organising and emergent processes, which we can simulate in the computer by writing codes that enable production in a manner whereby components self-configure. This is an approach which is becoming pervasive in schools of architecture and is influencing practice. Fig. 12.2,

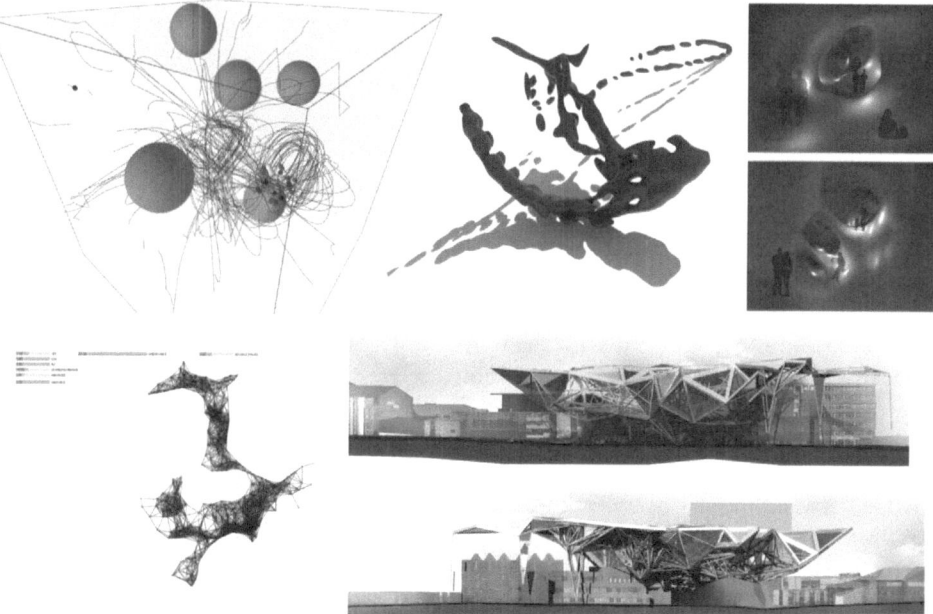

Figure 12.2 Top series illustrates the project *Paramecium Drawing*. On the top left, the agent-based model produced in Processing creates a primitive geometrical composition which is exported into Rhino (centre top). A mesh is wrapped around Processing results to form a composition which is rendered and enhanced in Photoshop to articulate a spatial scenario (top right). The bottom-left image illustrates the agent-based flocking model, which is then induced through the same process, fabricating final results (bottom right). Tim Ireland

for example, illustrates two projects by Leicester School of Architecture student Shen Guanlong (aka Jerry) from the Motive Ecologies program. The top project, *Paramecium Drawing*, derives inspiration from single-celled organisms, which are attracted to light. The organisms' behaviour was used as a mechanism to draw and generate spatial compositions that materialise bottom-up. The second project utilises the flocking of birds as a means to generate form through the interactions of a population of agents with a particular site, which freeze upon encountering certain conditions. The process involved writing bespoke code (in the Processing programming language)[26] to generate nascent results which are then exported into CAD software (AutoCAD and Rhino) to develop and tailor a final form, which are then enhanced graphically (using Photoshop) to express an architectural scenario. Beyond the nascent form generated through code, which constructs a population of autonomous entities which then navigate and negotiate "their" world to define some primitive outcome, the process relies on the dexterity of the designer and this is visually oriented. See Fig. 12.1. The methodology outlined is typical of the computational design production process because the way we engage with computers is fundamentally visual.

Architects have always looked to nature for inspiration. However, this has (throughout history) been a matter of analogy[27] and consequent to the visually oriented perspective that dominates. Frank Gehry's expressive and Calatrava's carcass-like buildings are indicative of an aesthetically oriented architecture that promotes the dominance of the visual over other architectural qualities, whereby the correlation between architecture and nature is determined by analogy. Philip Steadman explains that Gehry uses computational modelling software to develop his conceptual designs (which he creates through physical models) to model his shell structures in detail, which are then passed directly to computer-controlled manufacturing machines. This linking of CAD with computer manufacturing allows the generation of idiosyncratic "modular" forms, making it possible to assemble roofs or building envelopes whose organic-like forms resemble natural phenomena.[28] There is no reason, other than an overriding concern for the visual, that computational technologies that facilitate designing should resemble complex configurations. The aesthetic language that transpires is a misrepresentation and is based on the preoccupation of architects happier to play with shapes rather than the advancement of architecture through serious engagement with the innovations offered by the natural phenomena that inspired these formations. The ubiquitous use of computing technology in architectural design today remains largely entrenched in the dominance of sight principally because of the fact that digital technology is rooted in the capacity to display output graphically. Subsequently, interaction is conditioned by visual correspondence between input (mouse/pointer) and output (graphic display). The "eye-ball" test is implicit in architectural design, and whilst computation facilitates calculation of various parameters, this visual evaluation of results remains the principal criteria for determining success (see Fig. 12.1). As Paul Coates asks, "How do we observe the space and form without simplistic models of perception based on the eye as a camera and the mind as a logic machine?"

An incessant problem with computational designing is that what you put into the machine is what you get out. Techniques borrowed from computer science (that simulate natural processes) provide a means to generate shape and form, as well as produce solutions that correlate to the manifestation and generation of natural phenomena. These can surprise and entertain us, and they suggest creativity is occurring *in* the machine. There is, however, no (true) novelty produced in the process of production, because nothing new is inputted into the mix. For example, one might employ an evolutionary technique, such as the genetic algorithm (GA) to resolve some well-defined problem,[29] but GAs just shuffle parameters, so what transpires is combinatorial, arising from within the same possibility space. What is actually occurring is basically mixing, or what Peter Cariani refers to as 'combinatorial novelty'.[30] This is creative in so far as "it brings into being new combinations of elements. However, its use of fixed sets of primitive elements means that the set of possible combinations is closed".[31] As Bateson advises, a difference (something new) is required to affect a change, either added to the constituents implemented or the manner of their interaction. Importantly, this difference is not something we (the operator) input into the mix, but it is something that transpires as a product of the mix or the resulting interaction(s) between initial constituents. As Cariani argues, for "creative novelty" to emerge, a difference must be introduced into the system by external agents or processes. This is something inherent to natural systems because they are abductive. They are open to the environment and are thus susceptible to perturbations from outside their closed domain of relations.

Taking this to a fundamental level, the behaviour of two agents, one red and one blue, which move randomly may surprise the observer if (for instance) "red likes blue and blue dislikes red" produces a chasing behaviour, whereby blue evades red each time it encroaches on it. All that is required for this to occur is that the red one happens to take a step that ends up correlating to the position of blue, resulting in the occurrence that red is chasing blue, which is evading red. This banal example serves to illustrate how purposeful behaviour may arise but that in actuality it is no more than a product of coincidence. The relations are stipulated for this to arise and will never differ. The two agents will continue in perpetuity to either move randomly or imply pursuit. The blue will never become acquainted to red, or vice versa. What appears to be a novel behaviour is simply an outcome of fixed rules. The agents' relations need to be determined, or else the two agents do nothing but move randomly.

From *bio*logical to artificial

In *Out of Control*, Kevin Kelly argues a range of human undertakings, including design and manufacturing, are increasingly defined by "ways of biology", or what he calls "*bio*-logic". There are three forms of logic: deduction, induction and abduction. Deduction is about inquiring from the general to the particular: consisting of testing a general rule to fit or describe a particular case. Induction is the inverse. It infers a general rule from a particular case, proving that something *actually* operates. Abduction is the inference of the case from a general rule. It is hypothesising about the cause of an effect – the instance of which may be a surprise. Abduction is not accurate, but infers that something "may" be.

> Abduction is the process of forming an explanatory hypothesis. It is the only logical *operat*ion which introduces any new idea; for induction does nothing but determine a value and deduction merely evolves the necessary consequences of a pure hypothesis.[32]

Abduction is soft, a form of speculation, whereas deduction and induction are considered hard, because they are scientific modes of deriving facts from the world. Abduction is something "natural" in that facts are realised through acting in and observing phenomena from the standpoint of being in the world: a product of being embedded in an environment and thus consequent on organisms' joint sensorial capacities.

While separate methods of logical reasoning, the three forms are conjoined. However, abduction tends to be overlooked precisely because it is soft. Paul Bourgine and Franseco Varela emphasise the operational aspect of abduction as being productive and underlying what we perceive as purposive behaviour in natural systems.[33] Similarly, Lionel March surmises the usefulness of thinking of design in these terms: abduction (being productive) creates, deduction predicts and induction evaluates, thus forming a cyclic iterative process.[34] Outlining the significance of autonomy in their Introduction to the First European Conference on Artificial Life, Bourgine and Varela stipulate it is "clear that one central topic in artificial life is how to define and implement, as part of the systems closure, effective abduction processes".[35] The "*bio*logical" architecture of Kelly draws inspiration from nature, not by imitating its forms, but by understanding its behavioural traits.[36] To engage with the creative propensity of living systems, we need to engage with their abductive nature.

Chasing a machinic abduction

All living things dwell in behavioural niches, and in so doing, they affect their environment in some way. Furthermore, various organisms have developed the capacity to modify their environment in such a way that they construct artefacts. These structures embody the subject's intelligence. The web defines that of the spider, a bird's nests the predisposition of the species, the cathedral nest built by termites that of the colony and a building that of humans. While human beings may be understood to create artefacts "par excellence", their constructs are ingrained by patterns of inhabitation, i.e., the tendencies and nuances of living beings materialise, becoming transcribed in the artefacts they produce through their activities and capacity to affect matter. However, human constructs are conflated with abstract notions, ideals and agendas. The consequence of what Lefebvre called "abstract space". Whilst analysis and study of location and context enables designers to respond to a given site, they are not embedded in the same manner as a colony of ants or termites is (for example), thus they respond to differences and perturbations as they occur throughout the process of constructing their nests: an incessant process, which ends with the demise of the colony. Such constructs are situated by being a product of, and consequent to, the environment and the course of interaction between that which performs the fabricating and its environment. The process of construction is a consequence of environmental changes that feedback on the builder(s) to effect further changes. In other words, there is reciprocality between what is produced and its production, which is a process Pierre-Paul Grasse termed 'stigmergy': a kind of indirect, environmentally mediated communication whereby partially built structures direct current building action.[37] In other words (with regards to our ant and termite example), only those insects in proximity to some cue, and thus nearby in space-time, will be impinged upon and steered to act in some way. Such systems are abductive because they are open to the environment and are thus susceptible to perturbations from outside their closed domain of relations, which is something standard computation is incapable of replicating because (everyday) computing is deterministic.[38] The modern computer (based on the Turing machine) is fundamentally a tool to provide concrete answers to well-defined problems. It is not a tool for speculation but for resolving measurable problems. However, because the computer provides the capacity to work through problems quickly (enabling a method of working that employs an army of clerks),[39] it is utilised in this way. Consequently, a designer can test and work through connotations quickly and efficiently, but because of the analytical basis of the computer, it is not capable of speculation. Turing recognised this and proposed that for resolving immeasurable problems, an oracle needed to be coupled to his Turing machine. The oracle would provide the capacity to "think" abductively, like a shaman pre-empting future events by interpreting the signs.

Emphasising the social dimension of being in the world, Lefebvre stresses, "Space is social morphology: it is to lived experience what form itself is to the living organism, and just as intimately bound up with function and structure".[40] The manner in which something holds significance to some other, such as to effect a force, is intrinsic to sociality. That there is some *effect* between one thing and another means that these things enter into a relationship and thus have some form of commonality. The fact that it is the property of significance that brings this relation into being distinguishes this kind of semiotic causation from mere brute force causation, as the hallmark of relationships established by living beings.[41] We might consider that this effect has some value or that it is self-reinforcing such that it causes habit – a condition pertinent to the organism-in-its-environment and which extends up (in the evolutionary chain) from the primitive cell to human beings. "Social" thus infers some effect between two or more living beings (or at a more rudimentary level between an organism and those things it routinely interrelates with) and the mutually constitutive intersecting vectors of significance and that this effect is reinforcing.

Engaging technology at the level of the organism

Kiesler argued architects deal with forces not objects. Consequently, design is "not the circumscription of a solid but a deliberate polarization of natural forces towards a specific human purpose". He coined the phrase "Biotechnique" to emphasise this and to elaborate design as the influence of life in a desired direction. By engaging with the organism at the level of its habitual tendencies, architects and designers strive to capitalise on the creative (pattern-forming) capacities of the phenomena under scrutiny; because living things are, in principle semiotic, and in themselves, modelling systems.[42] Computational modelling enables a means to engage with the pattern-making tendencies of natural systems and is a means to put Kiesler's claim to practice. Doing so provides a means to explore possibilities, but the process relies on a fixed set of primitives and relations between them, meaning the systems potential remains within these set constraints. Subsequently, artists and designers are turning to organisms themselves to leverage their behaviour and bend their tendencies such that they are physically utilising the organism as the medium. Consequently, what these bio-technical tinkerers are doing is hacking into the organism system of sign relations, which they manipulate to bend and leverage their artefact-making tendencies. For example, Tomás Saraceno's arachnological compositions utilise the web fabricating tendencies of spiders to construct spatial compositions that reflect galaxy-like compositions (see Fig.12.3, left), and artist Ren Ri employs bees to construct other-worldly structures (see Fig. 12.4, right). Placing the queen bee in the middle of a box with worker bees building a natural beehive around her, Ren changes the position of the box every seven days according to the role of a dice to create a living object. Ollie Palmer evokes Argentine ants to choreograph *Ant Ballet* by meddling with the ants' navigation using an artificial pheromone that is applied with a robotic arm.[43] This is not a new approach. Villagers

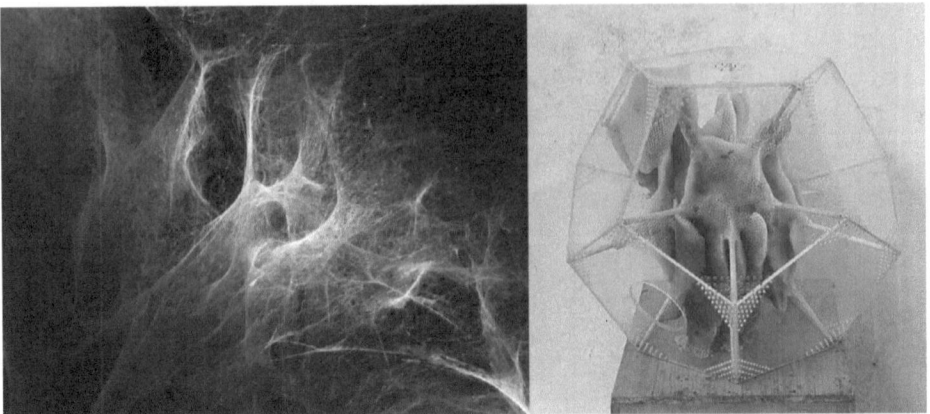

Figures 12.3 and 12.4 Tomás Saraceno arachnological composition (left), photographed by the author. Bee wax construction by Ren Ri (right)

in north-eastern India have been cultivating the roots of rubber trees for hundreds of years to build bridges across gorges.

At another scale, architects are utilising the productive capacities and pattern-forming tendencies of natural systems and testing how they may be used as a form of enabling device to generate spatial formations and organisational structures such that these are incorporated into both the design and fabricating process. For example, Studio Eric Klarenbeek is exploring ways of 3D-printing living organisms, such as mycelium (the threadlike network of fungi), which is used as a living glue for binding organic waste. In bringing together machine and nature Klarenbeek seeks to create a new material that could be used to make products with a negative carbon footprint. The *MORPHS* (Mobile Reconfigurable PolyHerda) of William Bondin were inspired by slime mould. The structures, conceived to be playful interactive public edifices, sense their surroundings and deposit digital data into their environment to inform their path and subsequent actions. Like living organisms, the morphs are self-preserving. Being solar powered, they avoid shade and avoid water to protect their electronics. The Growing as Building Project takes on the challenges of population growth and the reduction of natural resources to explore the idea of a living architecture by focusing on dynamically growing architecture which can adapt to the environment and the needs of users in a process of constant evolution. Specifically, the team is looking at material systems, such as the ones generated by mycelium, and growth principles found in the self-organising, explorative growth nature of slime mould. Furthermore, metabolic systems are developed in which organisms such as algae and bacteria are integrated into semi-closed loops generating, depositing and recycling building materials. Francois Roche is an architectural protagonist activist. A series of recent architectural experiments utilise robotics in a manner that capitalises on the capacity of machines to execute repetitive and monotonous tasks with accuracy whilst embracing the potential chaos inherent in feedback and biological processes. This is done not only to explore the margin between natural and artificial conditions but also as an agential means to prompt and critique social conditions. The Soliloquies project explores the correlation between a robotic arm and participant action, whereby the robot's fabricating motion is affected by the latter so as to upset the robots sanitary movement. This process is extended in *concrete[i]land*, in which the reading of a book affects the trajectory of the nozzle to create a shelter for petrified books. Voice intensity and tonal quality influence the seismograph movement of the robot, thereby affecting the production process.

The significance of these examples is an approach and disposition towards space, which engages with the distributed capacities of the phenomena studied and thus moves beyond the visual – both spatially and aesthetically. Advocating a spatial model that extends the classical notion of space, these artists and architects articulate a view of architecture concomitant with a model of the world that is distributed, self-organising and emergent. The classical view of the world as ordered, linear controlled and primarily considered as a visual phenomenon is transgressed; it got us to the moon, and may get us to Mars, but tends to put humankind in opposition to nature. The fundamental difference is one's perspective – one which doesn't see the world as Alberti claimed God sees it, but as (for example) an ant sees it. Doing so inherently puts one in tune with the conditions of becoming and being. Coupling such a view with the old ordained perspective has the potential to enrich architecture and, fundamentally, raises one's view over the parapet of sight alone to the richness of the distributed sensorial capacities which together determine our biological being in the world.

Conclusion

Putting the geometrical dexterity of computational modelling to one side, computational architectural design is fundamentally concerned with the analysis and generation of form to enhance the designer's aptitude in proposing solutions to design problems. Whilst computation offers a means to generate solutions that are a consequence of distributed and dynamic processes, and thus extend traditional design methods, the practice is visually oriented because the technology is defined by graphic interface (see Fig. 12.1). The premise is that computation facilitates a bottom-up, distributed, parallel, systemic means to design that will unleash the designer from his/her characteristic top-down solipsism, whereby an observer observing the outcome of their production is constrained by the inescapable paradox of observing their box from within. Computational design modelling enables the designer to side step Schrödinger's cat by conveying (a degree of) authority to their models to better test, speculate and scrutinise design solutions and thereby respond to the subjective and objective criteria of a design problem and the affordances generated scenarios offer. However, whilst computation allows greater versatility, it relies on what is given (i.e., the data input),[44] which sets the search space and limits the output to the closed domain of relations.

On the basis that behaviour is a product of semiosis, this chapter argues that computation as a tool in design needs to be reconsidered to account for qualia and the abductive (non-anthropocentric and non-materialistic) nature of living systems.[45] In so doing, the ontological conception of space needs reorienting to its biological origins,[46] because biological-space underpins architectural-space. Natural phenomena are creative and the result of creative processes. Computational architectural design is often motivated by natural phenomena and processes, but this is often confined to inspiration. This tendency should extend down to the constraints and processes of inaction underlying behaviour as a means to drive and extend current computational design practice. For example, evolutionary techniques have been presented as a means to generate solutions to design problems, but the techniques employed do little more than shuffle the constituent elements. They abstract and decouple the "agent of change" from the environment responsible (in large) for its transformation. If the computer is to be used as a tool for designing, and thus create novel solutions, it needs to be used in such a way that abduction drives methodology in order to capitalise on the creative processes of living things. Otherwise, design computation is limited to shuffling.

Behaviour is fundamentally a product of communication and signification determined by feedback loops, sign-action and the circumstances of an agent's embedding in, and interaction with, its subsuming environment. The effect of this non-linear, parallel and distributed systemic assembly is the distinction of boundaries affected through the course of the agent's unfolding interaction with its environment. It needs to be acknowledged that whilst the computer may provide a means to simulate parallelism and to engage with decentralised processes observed in nature, it does not, or at least the models currently produced by the community do not, correlate with the distributed semiotic processes defining the tendencies and production of natural phenomena. The components constituting natural systems have relevance and their interactions are deictic, effecting meaning and direction. Our computational modelling needs to account for these properties if they are to account for the behaviour and creative disposition of living things. Artists and designers such Ren Ri and Tomás Saraceno have tapped into the abductive qualities of natural phenomena to effect artefacts that move beyond visually dominated technologies. They have done this, in large part, by engaging at the level of the organism to capitalise on their pattern-forming tendencies, which they have leveraged and bent to create novel spatial formations and organisational structures. Nature is a creative engine which has long been an inspiration for designers. In moving beyond the analogical tradition in our computational design processes, we need to engage with the proclivities of the systems that inspire us and lift our gaze above the parapet of our visually oriented technologies. There is no reason, other than an overriding concern for the visual, that computational technologies that facilitate designing should resemble complex configurations. Such misrepresentation is more the preoccupation of architects happier to play with shapes than the advancement of architecture through serious engagement with the abductive nature of biotic systems.

Notes

1 The reading of signs is of cultural significance to how societies perceive and deal with determining future events. A shaman may, for example, interpret the entrails of a dead animal to determine the future and a young lover may pluck the flowers of a daisy to determine true love. Palm reading, tea-leaf reading and astrological reading of the stars are other well-known manifestations of people turning to others with the sense to read the signs they themselves cannot perceive.

2 Caucasian skin is renowned for turning colour when submitted to sunlight. This reaction is the effect of the skins ability to detect sunlight in a manner which the standard five-sense mentality does not account for. Too much sun over extensive periods can result in our skins cells becoming cancerous. In short, our skin, or rather our skin cells, has the capacity to detect UV light. A capacity not typically associated with human cognition. This distance between our body's sensorial capacity and the dominant view of humans having five senses is a result of Cartesian ontology, whereby the mental (res cogitans) is distinct from the body (res extensa): both of which are attributed by God.

3 As a verb, sight relates to perceiving with one's eyes (i.e., "you're a sight for sore eyes", or "land ahoy!") and in a transitive sense is oriented to instruments (such as microscopes and telescopes) which enhance our visual aptitude. The invention of the Renaissance perspective, for example, provided a means by which, as Alberti claims in his *Treatise on Painting*, "to see the world as God sees it". The idea of space being that which prevailed as a consequence of the ability to represent space in a manner as conceived in the mind (which the projective techniques of the Renaissance enabled) concretised a mathematical conception of space, which was thereby coupled with the infinite conception of space beheld by God, as opposed to the finite view experienced by ourselves. Consequently, society is influenced by the visual and the perceived significance of the visual has dominated architectural endeavour.

4 Henri Lefebvre, *The Production of Space*, translated by Donald Nicholson-Smith (Oxford: Blackwell Publishers Ltd., 1995).

5 Ibid., 96.

6 Perspective drawing enabled a means to see the world as God sees it, fuelling a particular articulation and understanding of space. The camera extended this providing photorealistic imaging and the capacity to capture stills of movement, which coupled the capacity to visualise space with time. Film amalgamated space and time by allowing capture of movement in time, and the computer has enabled the production of simulations and dynamic modelling. The premise here is that computation continues the same dynamic.

7 Ross Bryant, "Architectural Renderings Now 'Indistinguishable from Photos' Says Leading Visual Artist." *Dezeen* (2013). Accessed 5 December 2015. http://www.dezeen.com/2013/10/20/peter-guthrie-on-hyper-realistic-visualisations/.

8 Marcus Fairs, "Virtual Reality Will be More Powerful Than Cocaine." *Dezeen* (2015). Accessed 5 December 2015. http://www.dezeen.com/2015/04/27/virtual-reality-architecture-more-powerful-cocaine-oculus-rift-ty-hedfan-olivier-demangel-ivr-nation/.

9 Juhanni Pallasmaa, *The Eyes of the Skin: Architecture of the Senses* (London: Academy Editions, 1996).

10 For example, see Steen Eiler Rasmussen, *Experiencing Architecture* (Cambridge, MA: The MIT Press, 1959) and Peter Zumthor, *Thinking Architecture* (Boston, MA: Birkhäuser, 2006).

11 For example, sensors can be used to detect movement, distance, speed, heat, pressure, light, chemicals, vibration and so forth. Essentially, if a difference can be determined, it can be measured.

12 Tim Ireland, "Naturalising Space," in *Perception in Architecture*, ed. Claudia Perrin and Miriam Mlecek (Newcastle Upon Tyne, UK: Cambridge Scholars Publishing, 2015), 38–46.

13 Redrawn by author from Terence Broughton, A. Tan and Paul Coates, "The Use of Genetic Programming in Exploring 3D Design Worlds," in *CAAD Futures 1997: Proceedings of the 7th International Conference on Computer Aided Architectural Design Futures*, ed. Richard Junge (Dordrecht, London: Kluwer Academic Publishers, 1997), 885–917.

14 Parametricism is a stylistic ideology confined to producing discrete relational models.

15 Graham Nerlich, *The Shape of Space* (Cambridge: The Cambridge University Press, 1994), 4.

16 Bruno Latour, "Keynote" at *Topology Project: Spaces of Transformation* event at the Tate Modern Saturday, 3 March 2012. Part of "Continuity/Infinity: Keynote Conversation", with Olafur Eliasson, Bruno Latour and Peter Weibel. Chaired by Catherine Malabou.

17 In a collection of papers on the subject of space and geometry, Norbert Wiener (1915) argues that "Geometry is the science of a 'form' into which we cast our spatial experiences" (p95), claiming space is experiential and that geometry is an abstraction of that experience to determine a set of rules by which experience may be replicated. "Geometry is an *a priori* science, which deals with a certain schematization of experience, which we may call space [. . .] This schematization has a superficial appearance quite different from that of the experiences of which it is composed before they are schematized. [. . .] Experience never gives us a perfectly straight line, nor a precise circle, nor an absolutely accurate sphere. All these things, however, form topics dealt with in geometry. [. . .] a schematization of experience, not in the sense that it is a kind of approximate copy of experience with all the roughnesses left out, but in the sense that it is formed from experience by the application of some principle" (p. 99).

18 Berkeley claimed we have no direct access to reality (i.e., the external world), other than through our senses.

19 George Berkeley, *Oxford World's Classics: Principles of Human Knowledge and Three Dialogues*. Howard Robinson (ed.) (Oxford: Oxford University Press, 1996), 76.

20 Peter Weibel, *Beyond Art: A Third Culture: A Comparative Study in Cultures, Art and Science in 20th Century Austria and Hungary* (Wien: Springer-Verlag, 1997), 12–153.

21 Jacob von Uexküll, *Theoretical Biology* (New York: Harcourt, Brace & Company Inc., 1926), 9.

22 Gregory Bateson, "Form, Substance and Difference," in *Steps to an Ecology of Mind*, ed. Gregory Bateson (Chicago: The University of Chicago Press, 1972), 454–471.

23 William Mitchell, *The Logic of Architecture: Design, Computation and Cognition* (London: MIT Press, 1998), 1.

24 See Roberto Casati and Achille Varzi, *Parts and Places: The Structures of Spatial Representation* (Cambridge, MA: MIT Press, 1999).

25 Frederick Kiesler, "On Correalism and Biotechnique: A Definition and Test of a New Approach to Building Design," *Architectural Record* 86:3 (September 1939): 67.

26 Processing is an open source programming language and integrated development environment (IDE) built for the electronic arts, new media art and visual design communities. See https://processing.org/

27 Phillip Steadman, *The Evolution of Designs: Biological Analogy in Architecture and the Applied Arts* (Revised edition) (London: Routledge, 2008).

28 Ibid., 258–259.

29 Such as building envelope design. See Terence Broughton, A. Tan and Paul Coates, "The Use of Genetic Programming in Exploring 3D Design Worlds," in *CAAD Futures 1997: Proceedings of the 7th International Conference on Computer Aided Architectural Design Futures*, ed. Richard Junge (Dordrecht, London: Kluwer Academic Publishers, 1997): 885–917; and John Frazer, Julia M. Frazer, Xiyu Liu, Ming Xi Tang and Patrick Janssen, "Generative and Evolutionary Techniques for Building Envelope Design," in *Proceedings of 5th Generative Art Conference* 11th–13th December 2002, Italy, Milan, ed. Celestino Soddu, 311–316.

30 Peter Cariani, "Emergence and Creativity," *Emocao Art.ficial 4.0: Emergencia [Exhibition volume]*. Itau Cultural, Sao Paulo, Brazil (2009), 20–42.

31 Ibid., 2008, 7 of downloaded version.

32 Charles Sanders Peirce, "The Nature of Meaning," in *The Essential Peirce: Selected Philosophical Writings, Volume 2 (1893–1913)*, ed. the Peirce Edition Project (Bloomington: Indiana University Press, 1998), 216.

33 Paul Bourgine and Francisco Varela, "Introduction," in *Toward a Practice of Autonomous Systems: Proceedings of the First European Conference on Artificial Life*, ed. Francisco J. Varela and Paul Bourgine (Cambridge, MA: MIT Press, 1992), xi–xvii.

34 Lionel March, *The Architecture of Form* (Cambridge: The Cambridge University Press, 1976), 1–40.

35 Bourgine and Varela, "Introduction," XIV.

36 Kevin Kelly, *Out of Control: The New Biology of Machines, Social Systems, and the Economic World* (Cambridge: Perseus Books, 1994).

37 For example, see Dan Ladley and Seth Bullock, "The Role of Logistic Constraints on Termite Construction of Chambers and Tunnels," *Journal of Theoretical Biology* 234:4 (2005): 551–564; Guy Theraulaz, J. Gautrais, S. Camazine and J. L. Deneubourg, "The Formation of Spatial Patterns in Social Insects: From Simple Behaviours to Complex Structures," *Philosophical Transactions of the Royal Society A* 361(1807) (2003): 1263–1282.

38 See Emmanouil Zaroukas and Tim Ireland, "Actuating (Auto)Poiesis," *Systema Special Issue: Architectural Ecologies: Code, Culture and Technology*, ed. Liss C. Werner, 3:2 (2015): 34–55.

39 See Terence Broughton, A. Tan and Paul Coates, "The Use of Genetic Programming in Exploring 3D Design Worlds," in *CAAD Futures 1997: Proceedings of the 7th International Conference on Computer Aided Architectural Design Futures*, ed. Richard Junge (Dordrecht, London: Kluwer Academic Publishers, 1997), 885–917.

40 Henri Lefebvre, *The Production of Space*, translated by Donald Nicholson-Smith (Oxford: Blackwell Publishers Ltd., 1995), 94.

41 See Jesper Hoffmeyer, "Semiotic Scaffolding of Living Systems," in *Introduction to Biose-miotics: The New Biological Synthesis*, ed. Marcello Barbieri (Dordrecht: Springer, 2007), 149–166.
42 See Kalevi Kull, "A Semiotic Theory of Life: Lotman's Principles of the Universe of the Mind," *Green Letters: Studies in Ecocriticism: Special Issue: Biosemiotics and Culture* 19:3 (2015): 255–266.
43 Ollie Palmer, "Colony Choreography," *WIRED Magazine* 17 October 2012. Accessed 7 March 2016. http://www.wired.co.uk/magazine/archive/2012/10/start/colony-choreography
44 Which tends to hinge on what we know: that we tend to extrapolate from existing, or past, building scenarios, which leads to replication or at best reconfiguration of that which has been surveyed.
45 See Claus Emmeche, "The Computational Notion of Life," *Theoria – Segunda Epoca* 9:21 (1994): 1–30.
46 See Jacob von Uexküll, *Theoretical Biology* (New York: Harcourt, Brace & Company Inc., 1926), 1–51; and Tim Ireland, "Naturalising Space," in *Perception in Architecture*, ed. Claudia Perrin and Miriam Mlecek (Newcastle Upon Tyne: Cambridge Scholars Publishing, 2015), 38–46.

Bibliography

Bateson, Gregory. "Form, Substance and Difference." In *Steps to an Ecology of Mind*. Gregory Bateson (ed.), 454–471. Chicago: The University of Chicago Press, 1972.
Berkeley, George. *Oxford World's Classics: Principles of Human Knowledge and Three Dialogues*. Howard Robinson (ed.). Oxford: Oxford University Press, 1996.
Bourgine, Paul and Varela, Francisco J. "Introduction." In *Toward a Practice of Autonomous Systems: Proceedings of the First European Conference on Artificial Life*. Francisco J. Varela and Paul Bourgine (eds.), xi–xvii. Cambridge MA: MIT Press, 1992.
Broughton, Terence, Tan, A. and Coates, Paul. "The Use of Genetic Programming in Exploring 3D Design Worlds." In *CAAD Futures 1997: Proceedings of the 7th International Conference on Computer Aided Architectural Design Futures Held in Munich, Germany, 4–6 August 1997*. Richard Junge (ed.), 885–917. Dordrecht, London: Kluwer Academic Publishers, 1997.
Bryant, Ross. "Architectural Renderings Now 'Indistinguishable from Photos' Says Leading Visual Artist." *Dezeen* (2013). Accessed 5 December 2015. http://www.dezeen.com/2013/10/20/peter-guthrie-on-hyper-realistic-visualisations/.
Cariani, Peter. "Emergence and Creativity." *Emocao Art.ficial 4.0: Emergencia [Exhibition Volume]*. Itau Cultural, Sao Paulo, Brazil (2009). 20–42. Accessed 6 December 2012. http://www.cariani.com/CarianiNewWebsite/Publications_files/CarianiItauCultural2008-Emergence.pdf.
Casati, Roberto and Varzi, Achille C. *Parts and Places: The Structures of Spatial Representation*. Cambridge, MA: MIT Press, 1999.
Coates, Paul. *Programming Architecture*. London: Routledge, 2010.
Emmeche, Claus. "The Computational Notion of Life." *Theoria – Segunda Epoca* 9:21 (1994): 1–30.
Fairs, Marcus. "Virtual Reality Will Be More Powerful Than Cocaine." *Dezeen* (2015). Accessed 5 December 2015. http://www.dezeen.com/2015/04/27/virtual-reality-architecture-more-powerful-cocaine-oculus-rift-ty-hedfan-olivier-demangel-ivr-nation/.
Frazer, John H., Frazer, Julia M., Liu, Xiyu, Tang, Ming Xi and Janssen, Patrick. "Generative and Evolutionary Techniques for Building Envelope Design." In *Proceedings of 5th International Generative Art Conference GA2002*, 11–13 December 2002, Italy, Milan, 311–316.
Green, Keith. "The 'Bio-logic' of Architecture". *Proceedings for the 2005 ACSA National Conference* (2005): 522–530.

Hoffmeyer, Jesper. "Semiotic Scaffolding of Living Systems". In *Introduction to Biosemiotics: The New Biological Synthesis*. Marcello Barbieri (ed.), 149–166. Dordrecht: Springer, 2007.

Ireland, Tim. "Naturalising Space". In *Perception in Architecture*. Claudia Perrin and Miriam Mlecek (eds.), 38–46. Newcastle Upon Tyne: Cambridge Scholars Publishing, 2015.

Kelly, Kevin. *Out of Control: The New Biology of Machines, Social Systems, and the Economic World*. Cambridge: Perseus Books, 1994.

Kiesler, Frederick. "On Correalism and Biotechnique: A Definition and Test of a New Approach to Building Design." *Architectural Record* 86:3 (September 1939): 60–75.

Kull, Kalevi. "A Semiotic Theory of Life: Lotman's Principles of the Universe of the Mind." *Green Letters: Studies in Ecocriticism: Special Issue: Biosemiotics and Culture* 19:3 (2015): 255–266.

Ladley, Dan and Bullock, Seth. "The Role of Logistic Constraints on Termite Construction of Chambers and Tunnels." *Journal of Theoretical Biology* 234:4 (2005): 551–564.

Latour, Bruno. "Keynote" at *Topology project: Spaces of Transformation* event at the Tate Modern Saturday 3 March 2012. Part of "Continuity/Infinity: Keynote conversation", with Olafur Eliasson, Bruno Latour and Peter Weibel. Chaired by Catherine Malabou.

Lefebvre, Henri. *The Production of Space*. Donald Nicholson-Smith (trans.). Oxford: Blackwell Publishers Ltd., 1995.

March, Lionel. *The Architecture of Form*. Cambridge: The Cambridge University Press, 1976.

Mitchell, William J. *The Logic of Architecture: Design, Computation and Cognition*. London: MIT Press, 1998.

Nerlich, Graham. *The Shape of Space*. Cambridge: The Cambridge University Press, 1994.

Pallasmaa, Juhanni. *The Eyes of the Skin: Architecture of the Senses*. London: Academy Editions, 1996.

Palmer, Ollie. "Colony Choreography." *WIRED Magazine* 17 October 2012. Accessed 7 March 2016. http://www.wired.co.uk/magazine/archive/2012/10/start/colony-choreography

Peirce, Charles Sanders. "The Nature of Meaning." In *The Essential Peirce: Selected Philosophical Writings, Volume 2* (1893–1913). The Peirce Edition Project (ed.), 208–225. Bloomington: Indiana University Press, 1998.

Rasmussen, Steen Eiler. *Experiencing Architecture*. Cambridge, MA: The MIT Press, 1959.

Steadman, Phillip. *The Evolution of Designs: Biological Analogy in Architecture and the Applied Arts* (Revised edition). London: Routledge, 2008.

Theraulaz, G., Gautrais, J., Camazine, S. and Deneubourg, J. L. "The Formation of Spatial Patterns in Social Insects: From Simple Behaviours to Complex Structures." *Philosophical Transactions of the Royal Society A* 361(1807) (2003): 1263–1282.

von Uexküll, Jacob. *Theoretical Biology*. Published by Kegan Paul, Trench, Trubner and Company Ltd. New York: Harcourt, Brace & Company Inc., 1926.

Weibel, Peter. *Beyond Art: A Third Culture: A Comparative Study in Cultures, Art and Science in twentieth century Austria and Hungary*. Wien: Springer-Verlag, 1997.

Wiener, Norbert. (1915). "The Relation of Space and Geometry to Experience." In *Collected Works with Commentaries*. Volume 1. Pesi Rustom Masani (ed.), 87–217. Cambridge, MA: MIT Press, 1976.

Zaroukas, Emmanouil and Ireland, Tim. "Actuating (Auto)Poiesis." *Systema: Special Issue: Architectural Ecologies: Code, Culture and Technology*. Liss C. Werner (Guest ed.). 3:2 (2015): 34–55.

Zumthor, Peter. *Thinking Architecture*. Boston, MA: Birkhäuser, 2006.

Index

Note: figures and tables are denoted with italicized page numbers; endnote information is denoted with an n and note number following the page number.